Robt Holland   Monck Mason

Charles Green

AERONAUTICA;

OR,

SKETCHES

ILLUSTRATIVE OF

THE THEORY AND PRACTICE

OF

AEROSTATION:

COMPRISING AN ENLARGED ACCOUNT

OF THE LATE

AERIAL EXPEDITION TO GERMANY;

BY

MONCK MASON, Esq.

Member of the " Académie de l'Industrie française," and of the " Société de statistique universelle," etc. etc.

———

WITH PLATES.

———

Fredonia Books
Amsterdam, The Netherlands

Aeronautica:
The Theory and Practice of Aerostation

by
Monck Mason

ISBN: 1-4101-0453-2

Reprinted from the 1838 edition

Fredonia Books
Amsterdam, The Netherlands
http://www.fredoniabooks.com

In order to make original editions of historical works
available to scholars at an economical price, this
facsimile of the original edition of 1838 is
reproduced from the best available copy and has
been digitally enhanced to improve legibility, but the
text remains unaltered to retain historical
authenticity.

ROBERT HOLLOND, ESQ. M. P.

&c. &c. &c.

My Dear Sir,

If considerations of a public nature did not
sufficiently point out you as the proper subject of
the present dedication, reasons equally cogent,
though of a more personal character, might be
adduced in the very peculiar circumstances under
which our mutual acquaintance was first contracted and subsequently confirmed.

Various, indeed, and singular as are the occasions which have given rise to the social intimacies
of individuals, I question if any can be found
more unusual at once, and interesting to the
parties concerned, than that which first presented
us to each other's notice. While the generality of
mankind, in the ordinary routine of public or
domestic life, trace the first origin of their intercourse to the regions of the court, the camp,
the senate, or the bar, our acquaintance commenced in the car of a balloon; our *first* inter-
view was accomplished in the bosom of the sky;
our *second* was upon that memorable occasion

when, in concert with our expert and excellent friend, Mr. Charles Green, we performed the voyage, the narrative of which constitutes the principal feature of the following work.

To whom, therefore, could I with greater propriety dedicate the present volume, than to yourself, who have borne so conspicuous a share in the annals of the art it is professedly designed to illustrate? And now that you have quitted your station in the realms of empty space to take your seat among the councillors of your country, may I not be allowed to feel a little gratification in the reflection that, however different may be our views in matters of political concern, you still continue to regard with equal interest the relation of an event in which we both mutually participated, nor yet have had occasion to regret the intercourse to which it has given rise?

With all sincerity and respect,

I remain, my dear Sir,

Your obliged Friend and faithful Servant, &c.

MONCK MASON.

London, April, 1838.

CONTENTS.

LIST OF PLATES.

ERRATA.

Page 5, line 15, *for* "onset" *read* outset.

— 8, — 10 from bottom, *for* "1823" *read* 1822.

— 151, — 4, *for* "1827" *read* 1828.

— 164, — 15, *for* "21" *read* 18.

— 217, — 10 from bottom, *for* "October 5th" *read* September 28th.

— 231, — 10, *for* "Lyons" *read* Montpellier.

— — — 11, *for* "native" *read* Professor.

— — — 4 from bottom, *dele* "of the court of France and."

N. B.—This error was occasioned by the hurry in which the letter was originally prepared for the paper in which it appeared, and was not perceived until after it had been transferred into the present work. There was no court in France at the time.

— 304, — 14, *for* "upwards of" *read* nearly.

— 323, — 27, *for* "two thousand and fifty" *read* two thousand and fifty-six.

₀ The following alterations and additions to the particulars given in the catalogue of aeronauts, Appendix C. were either accidentally pretermitted or have come to the author's knowledge since that part of the work was sent to press.

Article GOWARD, *for* "Goward" *read* Gower, and *insert* 26th November.

— GREEN, (James) *insert* November.

— GRIFFITHS, *for* "1823" *read* 1822, and *insert* 30th July.

— GRISOLLE, *insert* Paris, 29th June, 1826.

— HEMMING, *for* "May" *read* 11th July.

— HOLLOND, (Mr. Richard), *insert* 24th May.

— IBRAIM PACHA, *insert* Warsaw, 14th May, 1790.

— JEFFRAYS, *for* "1829" *read* 1827, and *insert* 3rd September.

— JEPSON, *for* "Jepson" *read* Gypeon, and *insert* 14th May.

— JONES, *for* "1829" *read* 1828, and *insert* 23rd October.

— KENNET, *insert* 12th May.

— KENT, *for* "Dr." *read* Mr. B. A. and *insert* 30th August.

— LAPORTE, *for* "9th" *read* 19th.

— LAURENCIN, *for* "9th" *read* 19th.

— L'EPINARD, *before* "August" *insert* 26th.

— L'HOEST, *before* "July" *insert* 18th.

— LIGNES, *for* "9th" *read* 19th.

— LIVINGSTONE, *for* "1821" *read* 1816.

— LOCKER, *for* "4th" *read* 5th.

Article LOYED, *for* "Loyed" *read* Lloyd, and *insert* 21st September.
— LUWOF, *insert* St. Petersburg, 18th July, 1803.
— LUZARCHES, *insert* Javelle, 6th October, 1785.
— MAISON, *insert* Paris, 2nd August, 1790.
— MAITRE, *insert* Chambery, 12th May, 1784.
— MALCOLM, *for* "1826" *read* 1828, and *insert* 29th July.
— MARGAT, *insert* Paris, 5th June, 1817.
— MARR, *for* "17th" *read* 16th.
— MARSHALL, (Worksop), *insert* July.
— MARSHALL, (Derby), *for* "1828" *read* 1829, and *insert* September
— MATTHEW, *for* "1828" *read* 1827, and *insert* 23rd October.
— MICHAUX, *insert* Vienna.
— MILNES, *for* "1828" *read* 1829, and *insert* 16th May.
Page 269, *insert* PRINCE PUCKLER MUSKAU, Berlin, September, 1817.
Article NOLLIN, *insert* Paris, 14th July, 1801.
— OYESTON, *for* "Oyeston" *read* Oyston, and *insert* 19th Sept. 1831
— PATRICK, *before* "August" *insert* 16th.
— PAULY, *for* "19th October, 1805" *read* 4th August, 1804.
— PILTAY, *for* "Piltay" *read* Pilté.
— POTOSKY, *insert* Warsaw, 14th May, 1790.
— PUYMAURIN, *insert* Javelle, 6th October, 1785.
— RADCLIFFE, *before* "October" *insert* 31st.
— RAMSHAY, *for* "12th October" *read* 29th September.
— REICHARD, (M.) *insert* Berlin, September, 1817.
— REICHARD, (Madame) *for* "Brussels, &c." *read* Konigsberg, 22nd January, 1812.
— RICHARDSON, *for* "1828" *read* 1829.
— RIVIERRE, *insert* Isle de France, 19th June, 1784.
— ROBERTSON, (Gaspard), *before* "July" *insert* 18th.
— ROBERTSON, (Madame), *insert* Vienna, 8th October, 1804.
Page 271, *insert* RICHEY, Mr. London, 14th August, 1826.
Article ROGER, *insert* 27th July.
Page 271, *insert* ROGERS, Mr. J. Leeds, 2nd November, 1837.
Article ROLENS, *insert* 31st May.
— TURNER, *for* "1832" *read* 1831.
— VIPOND, *for* "1835" *read* 1834, and *insert* August.
— WOODHOUSE, *for* "16th" *read* 15th.
— MUSGRAVE, *for* "Musgrave" *read* Musgrave Tunnacliffe.
— PEMBERTON, *for* "1821" *read* 1828.
— POOLF, *for* "1821" *read* 1828.
— REDMAN, *insert* London.

INTRODUCTION.

———

THE interest with which the public at all times appear
to have regarded the progress of Aerostation, and especially the very flattering concern which they have deigned
so unequivocally to express for the successful issue of
our late undertaking, have concurred in inducing me to
abandon the usual path of communication hitherto adopted
upon such occasions, and confirm me in the opinion that
some account, more accurate and detailed than is generally to be found in the columns of the public press,
might not prove unacceptable to those for whose sympathy and consideration we can never acknowledge ourselves sufficiently grateful. In this belief, which I hope
may not be deemed fallacious, I have seized the first
vacant moment since our descent to embody in the present form all those incidents and observations to which a
voyage so singular is so amply calculated to give rise.

It is true that many of these have already reached the public ear through the medium of the public press, while at the same time no doubt much of the interest which owes its origin to the uncertainty and supposed peril of such exploits must have already subsided in the knowledge of the result, and of the leading features, which our duty to the public made it imperative upon us immediately to divulge. It is not, however, in the mere issue, successful or unsuccessful, that the chief merit or importance of such an enterprise can alone be said to consist. Designed with a view to special ends, and undertaken for the sole purpose of ascertaining and establishing the efficacy of certain improvements in the art, from which most beneficial results *were*, and I am now happy to add *are*, most likely to accrue, it becomes no less an obligation to ourselves than to the world in general, to make them partakers in the knowledge of whatever of interesting or important either accompanied the progress of our expedition, or may justly be expected to attend the adoption of those discoveries, the merits of which it was our sole object in the present instance to confirm. Before entering upon these details it may not be irrelevant to add a few words concerning the progress and present state of aerostation, in order that the nature and importance of the improvements they are intended to display, may be more conveniently appreciated.

From the time of the first discovery of the properties

and power of the balloon,* up to a late period, (already a lapse of more than half a century), a variety of obstacles

* It may not be uninteresting to those concerned in the annals of aerostation, to mention that the widow of the celebrated Montgolfier, the first inventor of the balloon, to which his name continues to be attached, is at this present moment living in Paris; and, though in her eighty-second year, in the perfect enjoyment of all her faculties, ardent in the advancement of the art, and hospitable in the reception of those who cultivate it. I had the pleasure of dining at her table, since our arrival, and of hearing from her own lips many of those curious anecdotes illustrative of its origin and progress, which, indeed, appear at all times to have accompanied the first dawning of great and important discoveries.

At this moment, while in the act of sending the present edition to the press, I have received the following letter from Madame Montgolfier, in acknowledgement of the receipt of a copy of this narrative, which I have just published in French, and which I did myself the honor to present to her. Independent of the interest attached to the writer, the very remarkable style of its expression, and its relation to the subject in question, I have deemed it worthy of being submitted to the public, as affording a testimony in favour of our enterprize, the more valuable as proceeding from the native of a foreign country, and one whose peculiar connection with the art confers a superior degree of importance upon the declaration of her opinion.

" J'ai lu, Monsieur, avec beaucoup d'intérêt votre voyage aérien; personne avant vous n'avait occupé aussi longtemps les divers royaumes de l'atmosphére, sur lesquels jusqu'à présent, nous n'avons guère que des connaissances douteuses. Je crois, Monsieur, qu'il vous est reservé, ainsi qu'à Mr. Green, l'intrépide navigateur, de nous apprendre avec plus de certitude

apparently insurmountable continued to obstruct the progress, and paralyze the efforts of all who sought to render it obedient to the sway of human will, and subservient to the purposes of human life. The chief of these impediments consisted in the uncertainty and expense attending the process of inflation from the employment of hydrogen gas; the dangers considered inseparable from the practice of the art; the difficulties which hitherto have baffled all attempts to give a direction to the ungovernable mass; and the impossibility which all previous aeronauts have experienced of remaining in the air a sufficient time to ensure the attainment of a sufficient distance.

To remove these obstacles and reduce the aerial vehicle to a more certain issue, a vast extent of actual expe-

ce qui se passe dans cette région si mobile, qui jusqu'à vous, Messieurs, n'a été frequentée que pour frapper à sa porte.

"Il n'est point étonnant, Monsieur que j'accompagne de tous mes vœux vos voyages; ce sont les pensées de mon mari que je vois reproduire. Le succès a quelque chose pour moi de personel; car si cette découverte est une heureuse idée, ce n'est qu'en la rendant utile, qu'on y attachera de la gloire.

"Je vous eusse, Monsieur, bien plutôt remercié de votre envoye, mais j'ai été souffrante; c'est encore ce qui me prive d'accomplir le désir que j'avais d'aller voir Madame votre mère : J'espere d'être plus heureuse dans quelques jours.

"Agreez, Monsieur, l'assurance de ma parfaite considération.
"MONTGOLFIER ' ETIENNE.'
"12 *Janvier*. 1837."

rience, united to an intellect capable of turning it to a proper account was absolutely required; and it would be an act of much injustice were I not to declare, that it is to the combination of both these in the person of Mr. Charles Green, that we are indebted for the entire results of all that is beneficial in the practice, or novel in the theory of this, the most delightful and sublime of all sublunary enjoyments.

It was to him, and to his discovery of the applicability of coal gas to the purposes of inflation, that we owe the removal of the first of those impediments in practice, which till then had continued to weigh down with a leaden hand the efforts of the most indefatigable and expert, and had, in fact, bid fair to quench the incipient science in its very onset.

Up to the period of that discovery, the process of inflation was one, the expense of which was only to be equalled by its uncertainty: two, and sometimes even three days of watchful anxiety have been expended in the vain endeavours to procure a sufficiency of hydrogen to fill a balloon, from which, on account of its peculiar affinities, it continued to escape almost as fast as it was generated; during all which time the various casualties of wind and weather, the inevitable imperfections of a vast and cumbrous apparatus, and above all the enormous expense attending the operation were to be incurred and endured, for the sole purpose, and with the sole object of remaining for a few hours helplessly suspended in the air. Under such disadvantages all prospect of advancement in the art had speedily disappeared; and it was only by the timely intervention of Mr. Green's ingenious application that the art itself was saved from a premature extinction—Aerostation had gone to sleep, when, roused by this discovery she awoke to redoubled efforts, and rendered that, in the hands of the skilful, a profession and a profit, which had ever before been a matter of doubt, difficulty, and distress.

By the adoption of the means which such a discovery now places within the power of the aeronaut, the laborious exertions of two and three days have become the affair of as many hours; and that which formerly could not have been accomplished under a cost of two and three hundred pounds, reduced to a scale of expenditure so low as not even to merit his consideration—unless when circumstances and the absence of competition may have left him at the merciless discretion of some unreasonable association. In illustration of the truth of this fact, I need only mention, that out of above two hundred ascents which Mr. Green has hitherto executed upon the same principles, throughout almost all parts of the United Kingdom, a large portion have been effected *without any expense of inflation whatever*; the various companies having gratuitously offered him the necessary supply of gas. Lest this should appear to some ex-

travagant or impossible, I beg to observe, that in a
country where coal abounds, as with us, the process of
distillation, by means of which the gas is procured, so far
from deteriorating the value of the material employed,
augments it so much that the residue (the coke), is
capable of producing by its sale, a return that covers
both the original purchase of the coals and the wages of
the men engaged in the operation. It is, therefore, the
mere wear and tear of the machinery alone, and the
interest of the money required to erect it, that can be
really said to constitute the expense of this once so
expensive an undertaking.

Independent, however, of these advantages accruing
from the adoption of coal gas in preference to hydrogen
for the purposes of inflation, there are others of great
importance, one of which especially merits notice. I
allude to the superior facility with which the former is
retained in the balloon, owing both perhaps to the greater
subtilty of the particles of hydrogen, and the stronger
affinity which they exhibit for those of the surrounding
atmosphere. In a balloon sufficiently impervious to retain
its contents of coal gas, unaltered in quality or amount
for the space of six months, an equal quantity of hydrogen
could not be maintained in equal purity for more than
an equal number of weeks. It will be unnecessary to
dilate upon the inestimable advantages which this property of coal gas presents to aerostation; especially when

we regard the future prospects of the art, its probable employment in the performance of voyages of long duration, and the difficulty, nay, impossibility in most instances of procuring or maintaining a supply of this perishable commodity.

With respect to the next of those impediments, which in the opinion of mankind might have continued to oppose its adoption as an organ of general utility—I mean the danger usually considered as consequent upon the exercise of the art, much is not required to prove the fallacy of such fears: two hundred and twenty-eight ascents,* undertaken at all periods of the year, without one disappointment to the public, and without one solitary instance of fatal consequences, or even of an accident productive of disagreeable results, (except from the intervention of malice),† ought to be a sufficient proof of

* The amount of Mr. Green's public ascents, up to the present period.

† In an ascent from Cheltenham, in the year 1823, in which Mr. Green was accompanied by Mr. Griffiths, some malicious individual contrived partially to sever the ropes of the car in such a manner as not to be perceived before the balloon had quitted the ground, when receiving for the first time the whole weight of the contents, they suddenly gave way, not, however, before the parties had time to secure a painful and precarious attachment to the hoop. Lightened of its load, the balloon with frightful velocity immediately commenced its upward course, and ere Mr. Green could gain possession of the valve-string, which the

how little danger is to be apprehended in the practice of
aerostation, when, under the management of a skilful

first violence of the accident had placed beyond his reach,
attained an altitude of upwards of ten thousand feet. The
situation of the parties at this juncture is easier to be conceived
than described. Clinging to the hoop with desperate retention,
not daring to rely any portion of their weight upon the margin
of the car, that still remained suspended by a single cord beneath
their feet, lest that also might give way, and they should be deprived of their only remaining counterpoise, all they could do was
to resign themselves to chance, and endeavour to retain their hold
until the exhaustion of the gas should have determined the career
of the balloon. To complete the horrors of their situation, the
net-work, drawn awry by the awkward and unequal disposition
of the weight, began to break about the upper part of the
machine, mesh after mesh giving way, with a succession of
reports like those of a pistol, while through the opening thus
created, the balloon began rapidly to ooze out, and swelling as
it escaped beyond the fissure, presented the singular appearance
of a huge hour-glass floating in the upper regions of the sky.
After having continued for a considerable length of time in this
condition, every moment expecting to be precipitated to the
earth by the final detachment of the balloon, at length they
began slowly to descend. When they had arrived within about
a hundred feet from the ground, the event they had anticipated
at length occurred; the balloon rushing through the opening in
the net-work with a tremendous explosion suddenly made its
escape, and they fell to the ground in a state from which with
great difficulty and after a long time they were eventually
recovered. Neither the author of this premeditated villany,
nor the design it was intended to answer, have ever yet been
discovered, although a reward of one hundred guineas was
immediately offered for his detection.

leader, and with the aid of those improvements to which his experience has given rise. It is not from the bungling efforts of unqualified persons that any judgment should be formed on this or other matters of practical detail; and where that skill is present, without which no one has a right to expect success, and those precautions have been observed which experience has shewn to be requisite, I do not hesitate to say, that the conveyance by the balloon is as devoid of extraordinary danger as that by any other mode of transport hitherto adopted.

To many persons, I am aware that this statement may appear rather exaggerated; and that, referring more to their own conviction than to the real nature and circumstances of the art, they may feel inclined to imagine that in the above description I have considerably underrated the difficulties as well as the dangers with which its practice is beset. Before, however, coming to so decided a conclusion, it would be well to inquire, how much of those convictions are founded upon facts, how much may be owing to collateral impressions, entirely independent of the real merits of the case. We all know to what an extent, the mere unwontedness of any occurrence is calculated to excite our apprehensions of its safety; it is not in aerostation alone, that the ancient adage, of "omne ignotum pro terribile" has been found completely to supersede the exercise of the judgment even of the most profound.

To these impressions, no doubt, the influence of example

has largely contributed. So much more tenacious is the mind, of whatever partakes of the horrible and disastrous, than of that which is merely distinguished by its brilliancy and good fortune, that the relation of one fatal accident is sufficient to tinge with the character of peril, a whole career of unequivocal success. In the present question this is particularly unjustifiable, as a very slight consideration of the case, is all that is required to demonstrate the error of the grounds, upon which such an opinion has been founded,—suffice it to say that out of above a thousand ascents which have hitherto taken place, eight alone have led to a fatal termination. Of this number five have arisen from the employment of fire, either as an agent in the ascent, as under the original system of the *Montgolfiere* or Fire-balloon, or else from its accidental introduction in the form of artificial fireworks, as in the case of the late Madame Blanchard. Three fatal accidents therefore are all that have hitherto occurred, to signalise the progress of an art, already but in its infancy; and of these not *one* has taken place under circumstances which come within the reservation to which I have above confined myself.*

* For the satisfaction of those who may wish to enquire more particularly into this subject, I have subjoined at the end of the narrative (*vide Appendix* c), a list of all those who have hitherto made personal trial of the art; together with a short account of such among them as have fallen victims to the pursuit, and of the circumstances to which their misfortunes are immediately to be attributed.

Another obstacle, which has hitherto most effectually opposed the adoption of aerostation, as an art subservient to the ordinary purposes of human life, is the want of means to guide the balloon according to a given direction. As the discussion, however, of this question would lead to a considerable digression, and as, moreover, it formed no part of the project in pursuance of which our late expedition was undertaken, I shall omit the mention of it for the present,* and pass at once to the consideration of those means whereby Mr. Green has succeeded in enabling the aeronaut to maintain the power † of his balloon undi-

* To enable the reader to form some opinion for himself upon the state of a question which has never ceased to interest, and indeed almost to agitate the speculative world ever since the first discovery of aerostation as an art, I have collected and thrown together the principal points in theory and practice which bear upon the matter of the guidance of the balloon according to a given course. If to no other purpose, it will at least serve to put persons on their guard, and enable them to defend themselves from the insinuating speculations of mistaken or designing projectors. *See Appendix* D *at the end of the volume*.

† The *power* of the balloon, in the technical language of aerostation, signifies the amount of the means whereby it maintains itself in the atmosphere. The measure of this amount can be expressed either in terms of her contents of gas, or of the burden they enable her to sustain. As the former, however, is of a variable value, depending upon its purity, the latter has been commonly adopted for the purpose, and the

minished during the continuance of the most protracted voyage it could ever be required to perform. In order fully to comprehend the value of this discovery, which more immediately formed the object of our late enterprise, it is necessary that some idea should be had of the difficulties it was intended to obviate, and of the effects they were calculated to produce upon the further progress of aerostation.

When a balloon ascends to navigate the atmosphere, independent of the losses occasioned by its own imperfections, an incessant waste of its resources in gas and ballast becomes the inevitable consequence of its situation. No sooner has it quitted the earth than it is immediately subjected to the influence of a variety of circumstances tending to create a difference in its weight; augmenting or diminishing, as the case may be, the power by means of which it is supported. The absorption, or evaporation of humidity, to an extent proportioned to its dimensions; the alternate heating and cooling of its gaseous contents by the remotion or interposition of clouds between the object itself and the influence of the solar rays, together with other more secret, though not less powerful agencies, all so combine

number of pounds avoirdupois required to her balance at the surface of the earth used to denominate the actual amount of her power.

to destroy the equilibrium which it is the main object of the aeronaut to preserve, that scarcely a moment passes without some call for his interference either to check the descent of the balloon by the rejection of ballast, or to controul its ascent by the proportionate discharge of gas; a process by which, it is unnecessary to observe, the whole power of the machine, however great its dimensions, must in time be exhausted, and sooner or later terminate its career by succumbing to the laws of terrestrial gravitation.

It is, however, at the two great epochs which mark the diurnal division of our time that those alterations take place, the effects of which are most seriously detrimental to the career of the balloon. Invested upon the approach of night with an accumulation of moisture, varying according to her size, from two to three hundred weight, the abandonment of an equal amount of ballast becomes absolutely necessary to enable her to maintain her position in the air. As the morning advances, and the rays of the sun begin to predominate, all this moisture becomes gradually dissipated, while incapable of repairing the loss of the weight which has thus surreptitiously superseded so large a portion of her own disposable stock, she becomes armed with a surplus of ascensive power, from which she is either liberated by the free exercise of the valve, or else effects that liberation for herself by the rapid elevation to which the incidental abstraction of so much weight has necessarily subjected her.* Another day ensues, and another night begins to require the repetition of the same proceeding. But no means now appear to satisfy the accruing demands; the gas which should have enabled her to support the extrinsic accumulation has already been dispersed, and the ballast which might have been rendered as an equivalent, the excesses of the preceding night have long since exhausted. In this dilemma no alternative remains, and the balloon, obedient to the impulse of her increasing preponderance, is gradually borne to the ground. Such is a rough, but correct, outline, of what must occur to every balloon under the ordinary circumstances of aerial navigation, and which has hitherto invariably confined the duration of all such excursions within the narrow limits of six-and-thirty hours.

As the real source of the injury inflicted upon the

---

* Lest it should not be quite apparent, how the mere change of altitude can avail to produce the effect here ascribed to it, it may be proper to observe, that as the pressure of the atmosphere upon the surface of the balloon becomes diminished in consequence of its ascent into a rarer medium, its gaseous contents proportionably expand, and filling whatever space may have been previously vacant, begin to escape through the neck of the balloon, which is at all times designedly left open, or otherwise are partially discharged by the action of the valve in anticipation of this result.

balloon in all these cases is the *permanent* abandonment of a portion of one of her resources to counteract the effects of a *temporary* deficiency in the relative power of the other, the natural remedy suggests itself in the disposal of those resources, *one or other of them* * after such a manner as will enable the aeronaut to recover again the portion expended, as soon as the circumstances which called for its employment shall have ceased to require its continuance. The mode in which this has been accomplished, according to Mr. Green's ingenious discovery, is as follows :—as soon as the balloon has sufficiently quitted

* The gas being liberated only in consequence of an occasional deficiency of ballast, and ballast being discharged only to compensate for an occasional deficiency of gas, it is evident that an effectual check being placed upon the accidental consumption of *either* must remove the contingency that caused the waste of *both*. Upon which of these resources, therefore, the gas or the ballast, it is designed to operate in effecting the remedy above-mentioned, is, theoretically speaking, a matter of indifference, as far as the *mere result is alone* concerned ; the end of either mode of proceeding being the same, viz. the maintenance of the original equilibrium, upon which depends the actual continuance of the balloon in her career. The application of the remedy, however, to the former supply appearing to be fraught with difficulty, and, indeed, long considered as unattainable by any means consistent with practical utility, it is the latter of these which Mr. Green has contrived to render subservient to the end proposed. For some observations touching the further extension of this remedial process, the reader is referred to the *Appendix* z at the end of the narrative, in which this matter is more fully discussed.

the earth, and circumstances appear to render it advisable, a rope, varying in length from a thousand feet upwards, according to the exigencies of the case, and of a mass proportioned to the weight against which it is intended to provide, is lowered from the car by means of a windlass, and passing through a pulley attached to the hoop above, is thus allowed to remain freely suspended in the air. As soon as any alteration takes place, whereby her specific gravity is increased, and the balloon in consequence begins to descend, the lower extremity of this rope becomes gradually deposited on the ground, and acting in this case like the discharge of so much ballast, keeps constantly abstracting from her weight below, in the direct proportion to the augmentation which it is receiving above, until the latter having reached its maximum, and an adequate compensation having been effected by means of the former, her further descent is eventually checked, and she either continues to advance upon the level to which these vicarious alternations have reduced her, or rising again under the influence of the first change that occurs, sufficient to produce such a tendency, and reversing in her ascent all the proceedings that attended her depression, she gradually becomes charged with all her former weight, and ultimately quits the earth in the same condition with regard to her resources in gas and ballast as she was ere circumstances had interfered to disturb the equilibrium of her previous course.

The value of this compensation, or in simpler phrase, the amount of the relief afforded to the balloon by means of the agency of the guide-rope, is to be found in the difference of force required to sustain a given weight and that required merely to overcome the resistance occasioned by its progress over the ground. Where, as in the case of a simple rope, there is no material impediment to its advance inherent in the form of the body itself, this difference is remarkably great; as was experimentally demonstrated by Mr. Green when he first conceived the idea of availing himself of such an application. A rope of one hundred and fifty feet in length, and weighing fifty-one pounds, when attached to a spring-beam and forcibly dragged over the uneven surface of the earth, indicated a resistance equal only to fourteen pounds; exhibiting even on this small scale a difference of about three-fourths of the entire weight: an amount considerably increased in the event of its adaptation to a balloon, where by the employment of a rope of larger dimensions, a much greater weight can be obtained with an equal or but slightly augmented power of resistance.* For all

* Where the length of the rope is determined, this resistance will be in the direct ratio of its weight, and *vice versa*. In the case of its application to a balloon, the weight is a given quantity, the resistance will therefore be in proportion to the length of the part constituting this weight, and consequently in favour of the heaviest, as being at the same time the shortest.

the purposes of aerostation this resistance is in fact of little or no moment; and even were it much greater than it really is, could scarcely be said to militate against the employment, or deteriorate from the merit of the discovery, when compared with the invaluable benefits it is otherwise calculated to confer; the more especially when we consider that, after all, the sole result of this resistance is merely a slight retardation in the rate of the balloon, accompanied by an equally slight depression in the level of her course from that at which she would otherwise have had to proceed.

The length of rope applicable to the above purpose, (beyond the quantity necessary to effect the requisite compensation) is a matter entirely depending upon the elevation at which circumstances may render it advantageous for the aeronaut to conduct his course. As far as mere security is concerned, a thousand feet, which was the amount we were provided with, is certainly sufficient to place him beyond the reach of any sudden emergency, though by no means enough to permit him to unite the advantages for which the guide-rope is especially intended, with those which an increase of elevation at the same time, might enable him to enjoy.* The occasions which render

---

* It is to the limited extent of our guide-rope that are to be attributed all those fluctuations to which we were *involuntarily* subjected during our late expedition, and which, without some

such an increase of elevation advisable, may proceed either
from the peculiar nature of the country over which he is
about to pass, or the desire to avail himself of the various
currents of air, which are known to exist at different altitudes in the atmosphere. Until we can assign the probable
limits of these occasions, it would be impossible exactly to
determine to what an extent it might not be advisable to
possess the means of increasing the altitude of the balloon
without entirely dissolving the connection with the earth,
afforded by means of the guide-rope. The possible extent
of these means, however, it is well to know, is only limited

explanation might, perhaps, appear to be at variance with what
has already been stated concerning its peculiar property of
maintaining the equilibrium of the balloon. The fact is, that a
thousand feet, the amount with which we were provided, is
much too small for general uses, and especially for the purposes
of a nocturnal voyage through countries the principal features
of which were but very imperfectly known to us. Once or
twice, therefore, during the night, induced by certain appearances
on the earth, or misled perhaps by the aspect of hills or precipices displayed at times by the vapours that hung above the
surface, we felt ourselves called upon to assume a higher level
than that to which the length of our guide-rope was at the time
confining us. On such occasions we were compelled to effect
by the discharge of ballast what we should otherwise have been
enabled to accomplish by a further descent of the rope, and
accordingly became exposed to the usual variations which
characterise all aeronautical voyages conducted under the usual
circumstances.

by the size of the balloon and the weight which, consequently, she is able to support. In a balloon, for instance, of the dimensions of that made use of by us upon the late occasion, had the whole of the ordinary ballast been sacrificed to the attainment of this one object, it would have been possible to have carried and employed a guide-rope of above fifteen thousand feet in length; an amount which the enlargement of the diameter of the balloon by less than one-fourth would have enabled us to double.

With regard to any objections which might have been urged on the score of the probability of the guide-rope becoming entangled or otherwise obstructed in its progress over the surface of the earth, the groundlessness of any such objections was originally ascertained by Mr. Green in several nocturnal ascents, executed in pursuance of his researches on this point, and since fully established by us in the present expedition, undertaken chiefly with a view to the determination of this very question. The fact is, that the guide-rope suffers little or no interruption to its progress whatever may be the nature of the country over which it may happen to be exercised. Trees, houses, rivers, mountains, valleys, precipices and plains, were all successively subjected to its course with equal security and indifference. A slight tremor in the car just sufficient to indicate the fact of its passage over an irregular surface, with occasionally a stronger vibration caused by

its momentary detention in surmounting some more prominent obstacle were the only perceptible effects that
accrued from its employment; and even from the latter of
these, a slight improvement in its construction suggested
at the time, would (as I have no doubt it will in future),
have sufficed to exempt us.

But, indeed, even if the guide-rope were liable to the
fullest extent of these objections, still no inference could
be fairly drawn from thence prejudicial to its employment,
inasmuch as an accident of the nature here contemplated,
should it occur, would not be attended with any serious
consequences, nor even with inconvenience beyond what
could be easily and promptly repaired. In the first place,
by the arrangement before described, the guide-rope being
virtually attached to the hoop, the whole strain of the
former in case of its seizure is thrown upon the latter
part of the machine, and the perpendicular position of
the car at all times effectually preserved. In the
next place, were such an occurrence to take place,
and the extremity of the guide-rope to become for-
tuitously arrested in its progress, nothing would be
simpler than by means of its appropriate machinery to
wind the balloon down to the object by which it was
detained, and either detach it from its hold or sever it at
once as close to the obstacle as the nature of the accident
would admit of its approach: at all events the utmost to
be apprehended, would be the loss of a portion of the

rope, more or less, according to the circumstances by which it had been occasioned.*

Thus, I hope, I have sufficiently explained the nature and mode of action of this simple discovery, so far as relates to its property of enabling the aeronaut to defend the equilibrium, and with it the power of his balloon,

---

* To anticipate any claims for priority which might happen hereafter to be set up with regard to this invention, it may be as well to observe, that several years have elapsed since Mr. Green first started the idea and determined its practicability by actual experiment; during which time he has been in the constant habit of availing himself of the indications which it will be seen it is calculated to afford in certain situations of difficulty.

About three years ago, an incident connected with the use of this guide-rope (then a secret), gave rise to a curious conclusion on the part of the simple inhabitants of a certain remote district in the county of Lincoln, and which, as it bears somewhat upon the present question, I take leave to relate here. An honest labourer in the fens, who happened to be afield rather early one fine morning, as Mr. Green was passing over-head, after an ascent he had made the previous night from the Royal Gardens, Vauxhall, seeing the guide-rope trailing on the ground, and totally unconscious of the speed with which it was proceeding, made an attempt to lay hold of it. No sooner, however, had it come into contact with his hand, than he instantly desisted, with an exclamation which became embodied in the report still prevalent in those parts, namely, that such was the velocity of Mr. Green's flight on the morning alluded to, that a rope which was hanging over the edge of the car had actually become *red-hot*, no doubt from the resistance created by the rapidity of its progress through the air !

against the various combinations which are incessantly tending to their abatement. To what an extent this saving agency might be capable of prolonging the career of the aerial voyager is a question entirely depending upon the condition of the balloon itself: where that is at its highest perfection, as in our particular case, I feel very little doubt that with no other drawbacks than those which may be expected to occur in the ordinary course of practice, the power of the balloon might be maintained sufficiently unimpaired to keep it afloat throughout the larger portion of a year.

The foregoing is not, however, the only advantage with which the employment of this ingenious application is attended. Two others of scarcely inferior importance remain to be noticed; I allude to the knowledge it enables the aeronaut to obtain with respect to the exact direction of the course he is pursuing, as well as of his distance from the surface of the soil,* at a time when, owing to the density of the night, the prevalence of fogs, the interference of clouds, or the absence of any assignable landmark (as in traversing an unbroken expanse of sea), such information would be otherwise altogether unattain-

---

* It is to be observed, that the indications of the barometer tend only to determine the altitude of the balloon, compared with a given level, usually that of the sea; with respect to its *immediate* distance from the surface of the subjacent soil it affords no information whatever.

able. In order to comprehend more fully the manner in which these indications are afforded, it is necessary to observe, that the progress of the guide-rope being delayed to a certain extent by its motion over the more solid plane of the earth's surface, while the movement of the balloon is as freely as ever controlled by the propelling action of the wind, the direction of the latter when in progress, must ever be in advance of the former; a comparison, therefore, of the relative positions of these two objects by means of the compass, must at all times indicate the exact direction of her course; while with equal certainty, an estimate can at once be obtained of the elevation at which she is proceeding, by observing the angle formed by the guide-rope, and a perpendicular let fall from the point of its attachment to the hoop. In proportion as this angle approaches a right angle, a diminution in the altitude of the balloon may be infallibly inferred; and, *vice versa*, its deviation from such will be found to correspond exactly with the increase of her distance from the earth. When the rope is dependent perpendicularly this deviation is at a maximum, and the machine, if in motion, may be considered as elevated to a height greater than the length of the guide-rope, or than that portion of it at the time employed.

The main feature, however, in this discovery, is the altered aspect under which it enables the aeronaut to regard the perils of the sea, and the consequent extension it bestows upon the hitherto limited sphere of his relations. Under the sway of such an instrument, the ocean, no longer the dreaded enemy of the aerial voyager, becomes at once his greatest friend; and instead of opposing his progress, offers him advantages more certain and efficacious than even the earth itself, with all its presumed security, is calculated to contribute. Freed from the apprehension of a forced determination to his career, he now regards in the sea but a vast plain ready to relieve him from impediments which might otherwise embarrass him in his course; in the ocean he beholds but a wider field for the exercise of those means which art has bestowed upon him to enable him to triumph over the difficulties of nature. In his view, the Atlantic is no more than a simple canal: three days might suffice to effect its passage. The very circumference of the globe is not beyond the scope of his expectations: in fifteen days and fifteen nights, transported by the trade winds, he does not despair to accomplish in his progress the great circle of the earth itself. Who now can fix a limit to his career?

P.S.—In one of the public journals (*Morning Chronicle, December* 21, 1836), which did me the honour to notice this little narrative upon its first appearance, objections were urged against some of the observations contained in the previous introduction, to which it becomes incumbent on me to advert. The first of these regards the degree of importance, so far beyond its merits, which I am stated to have attached to aerostation as an art applicable to useful purposes; the others refer in terms of question to the efficacy of the guide-rope as an instrument available to the end for which it is intended, and which I am said to have declared *"leaves nothing further to be desired"* for the complete perfection of the art. In answer to the first of these I beg to state, that far from representing faithfully my opinion on this point, the charge has no grounds whatever in any thing that appears in the pages of the work alluded to. After a careful revision of all the sentiments therein expressed, I cannot find that I have ever touched upon the question of its utility or adaptation to the ordinary purposes of human life, except where (*see page* 12), in enumerating the impediments that have hitherto obstructed its advancement, I have cited the existing impossibility of guiding the balloon according to a given direction as an effectual bar to the fulfilment of any such expectations. In all those passages in which, in allusion to Mr. Green's ingenious discovery, I have indulged in eulogies upon its merits, and attested, too energetically, perhaps, the benefits I considered it had entailed upon the art, it will be found upon reference that I have strictly alluded to the abstract qualities of *time* and *space*, the encreased *duration* of its powers, and the unbounded *extension* which has been consequently conferred upon its sphere of action.

Upon this point, I beg most distinctly to be understood; I am not one of those who put much faith in the applicability of the art to useful purposes, or who expect great results from any of the various schemes which are daily teeming from the brains of quacks or theorists on the subject of rendering the balloon a manageable engine, available to the wants and occasions of human life. As a source of recreation, at once the most interesting and sublime, it certainly stands unrivalled; in this sense it is, that I consider its present merit entirely to consist. What may be the case in future, when, (if ever it should come to pass), the guidance of the balloon to any serviceable extent shall have been effected, is another question, the probability of which, it would be as unbecoming in me to assert as in another to deny. In the meantime, all improvements in the art which tend to prepare it for this grand consummation, deserve, as I am sure they will obtain, the approbation and encouragement of all reflecting persons. Viewing them in this light it is, that I have humbly bestowed my modicum of applause and encomium upon the efforts of one amongst so many, who by his ingenuity and application has contrived to remove the dangers, diminish the expenses, and extend the relations of an art, which, however ultimately it might have been affected by the discovery of this grand arcanum, (the means of its direction) must without such advantages have still continued as inapplicable as ever to the ordinary purposes of human life.

With regard to the remaining objections which the critic of the Chronicle has thought proper to advance against the efficiency of the guide-rope, I beg to refer the reader to the concluding observations of the following narrative, and the general result of the expedition itself.

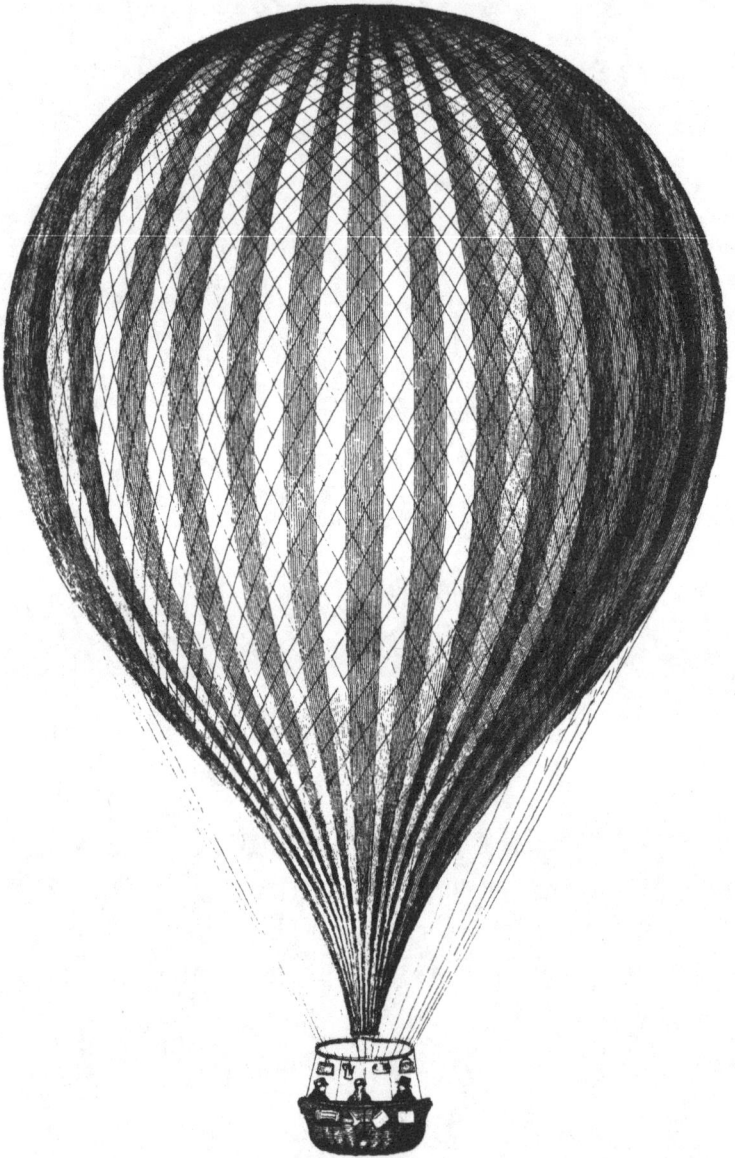

THE GREAT BALLOON.

From a Sketch by Robert Holland Esqre

F. C. Westley, 152, Piccadilly.

ACCOUNT, &c.

SUCH was the actual state of aerostation when Mr. Robert Hollond, a gentleman who had long cultivated a practical acquaintance with the art, resolved to afford an opportunity for a full display, and unequivocal determination of the merits of these discoveries, by undertaking at his own expense to fit out an expedition, under the guidance of Mr. Green, (in which he was so kind as to include me), for the purpose, and with the intention of starting from London and proceeding (in whatever direction the winds at the time prevailing might happen to convey us), to such a distance as would suffice to answer the ends for which the voyage was especially designed. In order to give the fullest effect to such an undertaking it was necessary to be provided with a balloon of size and structure superior to those employed upon ordinary occasions. Arrangements were accordingly entered into with Messrs. Gye and Hughes, the proprietors of Vauxhall Gardens, for the use of their large balloon, which they readily conceded,* at the same time placing their premises at our disposal for the purposes of the ascent. This balloon had been but recently built for them by Mr. Green, and combined in its construction all that the art and experience of the first aeronaut of the age could contribute to its perfection. In shape it somewhat resembles a pear; its upright or polar diameter exceeding the transverse or equatorial by about one-sixth; a form and proportion admitted to be at the same time most consistent with elegance of appearance, and most adapted to the wants and circumstances of aerostation. The silk of which it is formed is of the very best quality, spun, wove, and dyed expressly for the purpose; the utmost breadth of the gores, which are alternately white and crimson, is about forty-four inches; down the centre of each, and worked in the original fabric, runs a band or ridge of extra thickness, calculated to give additional strength to the texture of the material, and to arrest the progress of any rent or damage which might accidentally occur. The height of this enormous vessel is upwards of sixty feet; its breadth about fifty. When fully distended it is capable of containing

* It is but justice to the proprietors to state that no pecuniary consideration was required for the use of the balloon, which, together with the accommodation of their premises, was gratuitously tendered upon the occasion.

rather more than eighty-five thousand cubic feet of gas, and under ordinary circumstances is competent to raise about four thousand pounds, including its own weight and that of its accessories, which alone may be reckoned at about one-fourth of the above sum.

The car which appertains to this balloon is in proper keeping with its gigantic mate. It is composed of wicker-work, in the form of an oval, about nine feet in length and four in breadth. It is suspended by ten ropes to a hoop of six feet in diameter, and in thickness a like number of inches, formed of two circles of ash, one within the other, forcibly bent by steam, and retained in their position as well as strengthened by a triple tier of cable, which is enclosed between them. At either end of the car are two seats, fully capable of accommodating three persons each; while across it in the middle, and somewhat raised, is extended a bench about a foot in width, which, besides aiding to preserve the form of the vehicle against its own weight or other external pressure, served as a frame to support a windlass intended for the purpose of raising or lowering the guiderope whenever it should be required. In addition to these conveniences, the entire bottom of the car was on the present occasion fitted with a cushion, intended to be used as a bed in case adverse circumstances, by keeping us at sea or otherwise, should have compelled us to prolong the duration of our voyage to such an extent as to make it necessary to repose.

All the preliminary arrangements being now completed, after several unavoidable delays, occasioned chiefly by the weather, the day of departure was fixed for Monday, November 7, 1836, and the process of inflation having been commenced at an early hour, every thing was got ready for starting by one o'clock in the afternoon of the same day. As it had been resolved for special reasons that the ascent should not be made public, very few persons were present on the occasion within the precincts of the gardens. Outside, however, it was far different. Attracted by the prospect of the balloon during the process of its inflation, (no pains having been taken to conceal it from public view), a large concourse of persons had been assembling since an early hour in the morning, and by the time that all was completed, the multitude had already amounted to several thousands. So anxious, indeed, did they appear to witness the proceedings, that serious apprehensions began to be entertained towards the conclusion, lest the fences and palisades which enclosed the gardens might finally give way beneath the unwonted pressure of the numbers with which they were literally crowded.

The appearance which the balloon exhibited previous to the ascent, was, in truth, no less interesting than strange. Provisions, which had been calculated for a fortnight's consumption in case of emergency; ballast to the amount of upwards of a ton in weight, disposed in bags of different sizes, duly registered and marked, together with an unusual supply of cordage, implements, and other accessories to an aerial excursion, occupied the bottom of the car; while all around the hoop and elsewhere appended, hung cloaks, carpet-bags, barrels of wood and copper, speaking-trumpets, barometers, telescopes, lamps, wine jars and spirit flasks, with many other articles, designed to serve the purposes of a voyage to regions where, once forgotten, nothing could be again supplied.

Amongst the various contrivances which the peculiar circumstances of the case had led us to adopt, was a machine for warming coffee and other liquors, in which the heat developed in the process of slaking quick-lime was made to supersede the necessity of actual fire. This machine was found to answer the purpose perfectly well, although the dangers which it was intended to obviate are really not such as to require the aid of similar precautions. With that degree of prudence and attention which can at all times be commanded, no absolute peril need be apprehended from the employment of fire under proper restrictions. During the whole night we had a lamp burning constantly, nor did we at any time suffer anxiety on account of its presence, or perceive any occasion even temporarily to desire its extinction.

To provide against the inconveniences which we might have experienced subsequent to our descent, in continuing our journey through a foreign country, we likewise took

D

the precaution to furnish ourselves with passports directed to all parts of the continent, specifying the peculiar nature of our voyage, and entitling us to exemption from the usual formalities of office.

Finally, we were also charged with a letter to His Majesty the King of Holland, from Mr. May, His Majesty's Consul-General in London; which letter was put into the post-office at Coblentz, on the evening of the day succeeding our departure.*

---

* Of the due arrival of this letter, and his Majesty's gracious reception of it, we received the following testimony in a letter from Mr. May, shortly after we reached Paris.

London, November, 28, 1836.

Sir,

Perceiving from the accounts in the newspapers that you and your friends have arrived at Paris, I lose no time in having the satisfaction of thanking you very sincerely, for the care taken of the letter I took the liberty of entrusting to your kindness, for the purpose of having it forwarded to the King, at the Hague; it reached its destination on the 12th of November, through the post-office at Coblentz, and his Majesty was very much gratified at receiving a letter from England, by so novel a mode of conveyance as a balloon. The King has written a memorandum on the letter " *to be carefully preserved;*" wishing to keep it as a remembrance of this, as yet, extraordinary occurrence. I congratulate you and your companions on the success of your enterprize, and remain with great regard,

Sir,

Your most obedient humble servant,

J. W. MAY.

Thus prepared, and duly accoutred, at half-past one o'clock the balloon was dismissed from the ground, and rising gently under the influence of a moderate breeze bore speedily away towards the south-east, traversing in her course the cultivated plains of Kent, and passing in succession nearly over the towns of Eltham, Bromley, Footscray, and others, whose variegated outlines beautifully diversified the rich landscape that lay beneath us. The weather was uncommonly fine for the time of year; a few light clouds alone floated in the sky, and at least as useful as ornamental, served to indicate the existence of different currents at different altitudes; an information of which, it will be seen hereafter, we were enabled to avail ourselves with much effect.

Continuing in a south-easterly direction, at forty-eight minutes past two* we crossed the Medway, at the distance of about six miles to the west of Rochester, and in little more than an hour after† were in sight of the city of Canterbury, the lofty towers of its cathedral bearing distant about two miles, in an easterly direction. In

---

* The registry of the times and distances, as also of the direction of our course by the compass, during the voyage, was the particular province of Mr. Hollond, to whose Journal the author is indebted for all information on these points, as well as for many more valuable observations which will be found interspersed throughout the following narrative.

† Five minutes past four.

honour of the mayor and inhabitants of that city, under whose patronage our celebrated pilot had twice before ascended, we lowered a small parachute containing a letter addressed to the mayor, and couched in such terms as our hurried passage would permit us to indite.*

In a few minutes after † we obtained our first view of the sea, brightening under the last rays of a setting sun, and occupying the extreme verge of the horizon, in the direction in which we were now rapidly advancing.

During the latter period of this part of our voyage, the balloon, perhaps owing to the condensation occasioned by the approaching shades of evening, had been gradually diminishing her altitude, and for some time past had continued so near the earth as to permit us, without much exertion, to carry on a conversation with such of the inhabitants as happened to be in our immediate vicinity. So close indeed were we, at one time, as to be able distinctly to observe a covey of partridges, which either our approach or some other equally dreaded apparition had dislodged from their resting place, and sent to seek

---

* Of the due receipt of this letter, as well as of one to the same effect, which we subsequently addressed to the Mayor of Dover, we have since been informed; though the others which we discharged by similar means, at different periods of our voyage, we have reason to believe never reached the hands for which they were designed.

† Fifteen minutes past four.

a refuge on the borders of a wood which lay adjacent. A whole colony of rooks, alarmed no doubt by our formidable appearance, rose likewise in dismay, and after rending the air for miles round with their cries, and vainly trying the protection of the neighbouring woods, finally dispersed, scattering themselves in every direction over the surface of the earth beneath.*

Perhaps there is no situation conceivable from which the beauties of nature are seen to greater advantage, or with more singular effect, than that wherein the spectator is placed, when, seated in a balloon, he happens by circumstances to be brought into closer approximation with the earth beneath. The increased distinctness of the different objects, the novel aspect under which a vertical examination presents them to the view, the isolated position occupied by the beholder, and above all the exquisite motion, which however undistinguishable from its absolute effects upon the person, exhibits to the eye the ever-varying charms of rapid flight, are all effects perceivable under no other circumstances—and even denied to the aeronaut himself when occupying a higher range and indulging in a more extensive survey. It is not in fact the superior elevation or vast extent of prospect that under

* For some observations connected with the flight and disposition of birds in the air see *Appendix* F at the end of the narrative.

any circumstances constitute the real charms of such exhibitions, or contribute most to their enjoyment; and if we take the trouble carefully to examine the impressions which such scenes under such circumstances are wont to inspire, we shall find that to whatever class they may at first appear to be referrible, they are not nearly so much the offspring of pleasure as of surprise—of real critical delight as of that sort of gratification which is indebted to wonder and astonishment for its principal effect. To this conclusion I have been chiefly led by a consideration of the very beautiful appearance which the country presented, as under the influence of a gradual depression we slowly approached the ground, and for some time continued to skim along its surface at the slight elevation of a few hundred feet. The various objects, which, seen from on high, appeared like mimic representations of an ideal world, now gradually developed themselves and assumed the character and aspect of reality. The forests and parks, no longer an indefinite mass of something green, opened at our approach, separating into individual trees, the leaves and branches of which seemed almost within our grasp as we hurried over them. The houses, roads, enclosures, canals, and other minuter indications of civilized society, before scarcely appreciable, now also began to display themselves in their true colours, adding the charms of particular interest to that which was otherwise but imposing from its general effect; while the most interesting features of all, the living forms of nature, till now altogether invisible, began to mingle in the scene, and gave life and expression to what was otherwise at best but an inanimate, though brilliant, landscape.

About this time the first opportunity occurred of shewing how far it is possible for the skilful and experienced aeronaut to influence the course of his aerial vessel, by availing himself of the advantages which circumstances frequently place at his disposal. Shortly after we had lost sight of the city of Canterbury a considerable deviation appeared to have taken place in the direction of our route. Instead of pursuing our former line of south by east, which was that of the upper current, by means of which we had hitherto advanced, it became apparent that we were now rapidly bearing away upon one which tended considerably to the northward, and which, had we continued to remain within the limits of its influence, would have shortly brought us to sea, in the direction of the North Foreland. As it had all along been an object to proceed as near to Paris as circumstances would permit,* we resolved to recover as soon as possible the advantages which a superior

---

* The proprietors of the balloon having contemplated making an ascent from Paris, and Mr. Hollond having undertaken to transfer the balloon thither, it became a consideration with us not to increase our distance from that capital, more than was consistent with the main object of the expedition.

current had hitherto afforded us; and accordingly rose to resume a station upon our previous level. Nothing could exceed the beauty of this manœuvre, or the success with which the balloon acknowledged the influence of her former associate. Scarcely had the superfluous burden been discharged proportioned to the effect required, when slowly she arose, and sweeping majestically round the horizon, obedient to the double impulse of her increasing elevation and the gradual change of current, brought us successively in sight of all those objects which we had shortly before left retiring behind us, and in a few minutes placed us almost vertically over the Castle of Dover, in the exact line for crossing the straits between that town and Calais, where it is confined within its narrowest limits.*

Up to the present moment nothing had appeared cal-

---

* To the circumstances of this transaction, the apparent retardation of our course by the circuitousness of the route, the length of time we consequently remained in sight, and above all the rectilinear direction of our approach, is undoubtedly to be attributed the observation contained in the newspapers, that the progress of the balloon did not exceed the rate of four or five miles an hour; an assertion which a slight consideration of the time we had left London, and the distance we had accomplished, would have been sufficient to disprove. According to the above method of calculation, the mean rate of our course up to the time referred to, was somewhat more than twenty-five miles an hour.

culated to confer particular distinction upon our enterprise,
or to awaken the impression, that what we had undertaken
differed in any respect from the usual class of such excursions. The case, however, was now shortly to be
changed; a new and untried element was about to enter
upon the scene, producing new relations and requiring
the exercise of new resources. The knowledge that
whenever we might feel inclined it was in our power to
terminate our voyage by descent, (which gives such a
sense of security to all excursions over land), was about
to yield to the conviction that, no matter how urgent the
desire, how imperious the necessity, *that* expedient would
in future be withheld from us until it had pleased Providence to convey us to new regions, and afford us once
more the circumstances of a solid resting place. When
or where that might be, was a question as doubtful as the
winds by which alone it could be determined; nor was it
the smallest of the many charms peculiar to our situation,
that it was, and must for some time remain a matter of
the most complete incertitude what portion of the globe
was next destined to receive us. Confident, however, in
our own resources, I may safely assert that it was to us a
matter of the most perfect indifference in what manner
that uncertainty should be decided; and I feel convinced
that I but speak the sentiments of the whole party when
I declare that not a single particle of anxiety as to our
own personal safety for a moment disturbed the ardent

desire we all felt to push to a creditable bearing the enterprise in which we were embarked.

It was forty-eight minutes past four when the first line of waves breaking on the beach appeared beneath us, and we might be said to have fairly quitted the shores of our native soil, and entered upon the hitherto dreaded regions of the sea.

It would be impossible not to have been struck with the grandeur of the prospect at this particular moment of our voyage; the more especially as the approaching shades of night rendered it a matter of certainty that it would be the last in which earth would form a prominent feature, that we might expect to enjoy for a considerable lapse of time. Behind us, the whole line of English coast, its white cliffs melting into obscurity, appeared sparkling with the scattered lights, which every moment augmented, and among which the light-house of Dover formed a conspicuous feature, and for a long time served as a beacon whereby to calculate the direction of our course. On either side below us the interminable ocean spread its complicated tissue of waves without interruption or curtailment, except what arose from the impending darkness, and the limited extent of our own perceptions. Slightly agitated by a wind, unfelt by us, its pliant surface glistened faintly as it rose and fell, catching for an instant by the momentary obliquity of its parts the few rays of light that still lingered above the horizon, and losing them again as they turned their opposing outlines towards a darker quarter. On the opposite side a dense barrier of clouds rising from the ocean like a solid wall, fantastically surmounted, throughout its whole length, with a gigantic representation of parapets and turrets, batteries and bastions, and other features of mural fortification, appeared as if designed to bar our further progress, and completely obstructed all view of the shores towards which we were now rapidly drawing nigh. Upon the glittering plain which thus lay stretched before us, a few straggling vessels, some of which had already began to mount their lights, alone appeared, issuing from beneath the dark mantle of clouds that rested, as it were, upon the very bosom of the deep. In a few minutes after, we had entered within its dusky limits, and for a while became involved in the double obscurity of the surrounding vapours and of the gradual approach of night. Not a sound now reached our ears; the beating of the waves upon the British shores had already died away in silence, and from the ordinary effects of terrestrial agitation our present position had effectually excluded us.

I scarcely know whether it is an observation worthy of being committed to paper, but the sea, unless, *perhaps* under circumstances of the most extraordinary agitation, does not it in itself appear to be the parent of the slightest sound. Unopposed by any material obstacle, an awful stillness seems to reign over its motions. Nor do I think that even under *any* circumstances, no matter how violent, can any considerable disturbance arise from the conflict of its own opposing members. The impossibility of ever having been placed in a situation to bring this fact under the cognizance of our senses, is no doubt the reason why it has never before been noticed. On the shore or in the sea, no one has ever been present, independent of that material support, the absence of which is necessary to the success of the experiment; it is in the balloon alone, suspended in elastic ether, that such a phenomenon could either have been verified or observed.

According as we proceeded the lower strata of the vapoury bed in which we rested would slowly appear to dissolve, and opening beneath us, occasionally reveal a partial glimpse of the sea, now rapidly beginning to assume the sable livery of night. Across the field of view which thus became exposed, a solitary ship might now and then be seen to pass, entering at one side like the spectral representation in some magic lantern, and having sped its course, silently disappearing on the other. Wreaths of mist shortly after intervening, the whole would be swept from our view, leaving us once more enveloped in the dark folds of the prevailing vapours.

In this situation, we prepared to avail ourselves of those contrivances, the merits of which, as I have already stated, it was one of the main objects of our expedition to ascertain; and consequently, to provide against the increase of weight proceeding from the humidity of the atmosphere, naturally to be expected on the approach of night, we commenced lowering the guide-rope, with the floating ballast attached, which we had provided for the occasion.

Scarcely, however, had we completed our design, and were patiently awaiting the depression we had anticipated, ere the faint sound of the waves beating against the shore again returned upon our ears, and awakened our attention. The first impression which this event was calculated to convey, was that the wind had changed, and that we were in the act of returning to the shores we had so shortly before abandoned. A glance or two, however, served to show us the fallacy of this impression; the well-known lights of Calais and of the neighbouring shores were already glittering beneath us; the barrier of clouds which I have before mentioned as starting up so abruptly in our path, as abruptly terminated; and the whole adjacent coast of France, variegated with lights, and rife with all the nocturnal signs of population, burst at once upon our view. We had, in fact, crossed the sea; and in the short space of about one hour, from the time we had quitted the shores of England, were floating tranquilly, though rapidly, above those of our Gallic neighbour.

It was exactly fifty minutes past five when we had

thoroughly completed this *trajet*; the point at which we first crossed the French shore bearing distant about two miles to the westward of the main body of the lights of Calais, our altitude at the time being somewhat about three thousand feet above the level of the ocean. As it was now perfectly dark we lowered a Bengal light, at the end of a long cord, in order to signify our presence to the inhabitants below; shortly after, we had the satisfaction to hear the beating of drums, but whether on our account, or merely in performance of the usual routine of military duty, we were not at the time exactly able to determine.

Before dismissing the sea, a word or two seems required to counteract a vague and incorrect impression regarding its peculiar influence upon the buoyancy of the balloon, arising from the difficulties experienced by Messrs. Blanchard and Jeffries in their passage of the same straits in the year 1785, and the apparently unaccountable remotion of these difficulties as soon as they had reached the opposite coast. So many, however, are the circumstances within the range of aeronautical experience to which, without intruding upon the marvellous, or calling new affinities into existence, these effects can be satisfactorily attributed, that the actual difficulty lies in ascertaining to which of them they are most likely to have owed their origin. Of these the increase of weight by the deposition of humidity on the surface of the balloon,

occasioned by the colder atmosphere through which the
first part of their journey had to be pursued, and the subsequent evaporation of the same by the rise of temperature to which they necessarily became subjected as soon
as they came within the calorific influence of the land, is
in itself quite sufficient to explain the difference that
existed in the buoyancy of the balloon, during the different stages of its progress. Even in the absence of
any humidity whereby the actual weight of the balloon
could have been increased, the mere diminution of temperature, by condensing its gaseous contents, and their
subsequent rarefaction by the altered temperature they
were sure to encounter when they reached the opposite
coast, is more than enough to account for even much
greater effects than those to which it is here intended to
apply. As far as we were concerned certainly no such
uncommon impression was observable, nor did we experience any diminution of ascensive power in our transit
across the sea, beyond what we should have expected
under similar circumstances over a similar extent of land.

Having thus completed what may be termed the first
stage in our eventful voyage, we set about making such
preparations as the altered circumstances of the case
rendered advisable. For this purpose, the copper vessels
which had been intended to be used at sea if required, but
which our rapid passage over that element had left us no
opportunity of exercising, were again raised and with-

drawn, and a simple guide-rope of about a thousand feet in length substituted in their stead. Our lamp also was lighted, and so disposed as that in case of any appearance of danger, which, however, we neither anticipated nor experienced, it could be lowered instantly to a considerable distance from the car.* These arrangements being over, and nothing for the present appearing to demand our particular attention, we gladly availed ourselves of the opportunity to allay the cravings of an appetite which the incessant occupation of the previous day had hitherto prevented us from regarding. To this effect much preparation was not required. The bench, which

* Beyond the risk attendant upon the use of fire under ordinary circumstances there is but one situation peculiar to aerostation in which any particular danger is to be apprehended, or any particular precautions are necessary to be adopted: I mean when the balloon, in consequence of its elevation in the atmosphere, has become so much distended as to occasion the partial liberation of its contents. In such cases, which we experienced not unfrequently during the night, all that is required is merely to lower the lamp by means of a cord to such a distance from the car, as to place it beyond the reach of the gas issuing from the neck of the balloon. If it should be necessary to discharge gas from the valve, before this is accomplished care should be had to do so by degrees, not all at once, as the balloon being at such moments always in the act of rising would shortly enter into the atmosphere of gas thus created around it, which, if sufficiently impregnated would ignite, and most probably occasion the destruction of the machine.

we have before described as forming the central division of the car, served us most conveniently as a table, and was quickly spread with the good things which had been abundantly provided to cheer our solitary flight.   Cold meats of various kinds, beef, ham, fowl, and tongues, together with bread and biscuits, and a due admixture of wine and other liquors* formed the bases of a repast which might in truth have proved acceptable to much more fastidious palates than ours, especially tempered as they were by the rigorous discipline of a twelve hours fast, and a proportionate amount of bodily exertion.   Accordingly, with many a joke, touching the *high* flavour and *exalted* merits of our several viands, which, however agreeable under the circumstances, will not bear repeating here, we contrived to do ample justice to the good cheer, not forgetting amid the festivities of the scene, to drink a cordial health to the memory of those whom we had left

* For the benefit of such lovers of good cheer as may in future be tempted to prove the pleasures of aerostation, it may be as well to observe that it is not all liquors that can be conveniently employed upon such occasions.   Champagne, for instance, and bottled porter, cider, soda water, and all those which are generally termed " up in bottle," however anomalous it may appear, are by no means adapted for aerial excursions; their natural tendency to *flying* being so much accelerated by the diminished pressure which is the consequence of their elevation that they invariably *fly off* altogether, almost as soon as they have quitted the ground.

in uncertainty behind us. With an economy, however, which had in it somewhat peculiar, the bones and other fragments instead of being thrown over, were carefully collected in order to be employed for ballast whenever occasion might require. We also took the opportunity of proving the efficacy of our newly-invented machine for heating coffee, and found it answer the purpose fully as well as we had expected.

The night having now completely closed in, and no prospect of any assistance from the moon to facilitate our researches, it was only by means of the lights which either singly or in masses, appeared spreading in every direction, that we could hope to take any account of the nature of the country we were traversing, or form any opinion of the towns or villages which were continually becoming subjected to our view.

The scene itself was one which exceeds description. The whole plane of the earth's surface, for many and many a league around, as far and farther than the eye distinctly could embrace, seemed absolutely teeming with the scattered fires of a watchful population, and exhibited a starry spectacle below that almost rivalled in brilliancy the remoter lustre of the concave firmament above. Incessantly during the earlier portion of the night, ere the vigilant inhabitants had finally retired to rest, large sources of light, signifying the presence of some more extensive community, would appear just looming above the distant horizon in the direction in which we were advancing, bearing at first no faint resemblance to the effect produced by some vast conflagration, when seen from such a distance as to preclude the minute investigation of its details. By degrees, as we drew nigh, this confused mass of illumination would appear to increase in intensity, extending itself over a larger portion of the earth, and assuming a distincter form and a more imposing appearance, until at length, having attained a position from whence we could more immediately direct our view, it would gradually resolve itself into its parts, and shooting out into streets, or spreading into squares, present us with the most perfect model of a town, diminished only in size, according to the elevation from which we happened at the time to observe it.

It would be very difficult, if not impossible, to convey to the minds of the uninitiated any adequate idea of the stupendous effect which such an exhibition, under all its concomitant peculiarities, was calculated to create. That we were, by such a mode of conveyance, amid the vast solitude of the skies, in the dead of night, unknown and unnoticed, secretly and silently reviewing kingdoms, exploring territories, and surveying cities, in such rapid succession as scarcely to afford time for criticism or conjecture, was in itself a consideration sufficient to give sublimity to far less interesting scenes than those which formed the subject of our present contemplations. If to this be added the uncertainty that from henceforward began to pervade the whole of our course—an uncertainty that every moment increased as we proceeded deeper into the shades of night, and became further removed from those landmarks to which we might have referred in aid of our conjectures, clothing every thing with the dark mantle of mystery, and leaving us in doubt, more perplexing even than ignorance, as to where we were, whither we were proceeding, and what were the objects that so much attracted our attention—some faint idea may be formed of the peculiarity of our situation and of the impressions to which it naturally gave rise.

In this manner, and under the influence of these sentiments did we traverse with rapid strides a large and interesting portion of the European continent; embracing within our horizon an immense succession of towns and villages, whereof those which occurred during the earlier part of the night, the presence of their artificial illumination alone enabled us to distinguish.

Among these latter, one in particular, both from its own superior attractions, the length of time it continued within our view, and the uninterrupted prospect which our position directly above it, enabled us to command, captivated our attention and elicited constant expressions of admiration and surprise. Situated in the centre of a district which actually appeared to blaze with the innumerable fires wherewith it was studded in every

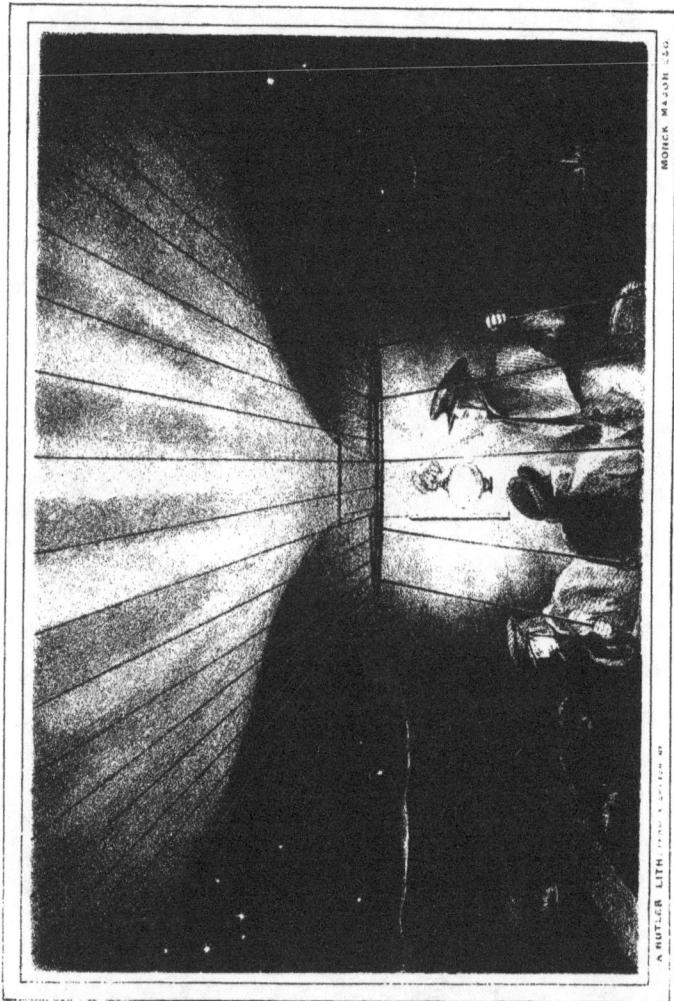

ENVIRONS OF LIÈGE, SEEN FROM THE BALLOON AT NIGHT.

direction to the full extent of all our visible horizon, it seemed to offer in itself, and at one glance, an epitome of all those charms which we had previously been observing in detail. The perfect correctness with which every line of street was marked out by its particular line of fires; the forms and positions of the more important features of the city, the theatres and squares, the markets and public buildings, indicated by the presence of the larger and more irregular accumulation of lights, added to the faint murmur of a busy population still actively engaged in the pursuits of pleasure or the avocations of gain, all together combined to form a picture which for singularity and effect certainly could never have been conceived. This was the city of Leige, remarkable from the extensive iron-works which, abounding in its neighbourhood, occasioned the peculiar appearance already described, and at the time led to that conjecture, concerning its identity, the truth of which a subsequent enquiry enabled us to confirm.

Almost immediately after we had passed the main body of the buildings, and before we had got quite clear of the outlets of the town, an accident deprived us of the use of our machine for heating coffee; just at the time too, when, from the increasing rigour of the night, its services were likely to prove most particularly acceptable. Previous to our arrival in the neighbourhood of so extensive an assemblage of buildings, we had thought it advisable to suspend the action of the guide-rope, by removing to such an elevation as would dissolve its connection with the earth, and carry it clear of the houses.* In this manner we had crossed the city, and were about to enter on the suburbs, when a slight tendency to depression made it necessary to discharge a small quantity of ballast

* It will very naturally be observed, that, having once submitted to interrupt the action of the guide-rope at a time when the original equilibrium of the balloon is under the influence of its greatest disturbance, (as, for instance, during the course of the night), by dissolving even for a moment its connection with the earth, (which is only to be effected by a rejection of ballast equal to the weight of rope upon the ground), all the advantages which had been previously obtained by the use of it are forfeited at once, and the aeronaut placed in exactly the same circumstances as if he had proceeded so far without the aid of such an instrument at all. This observation is essentially correct; nor would we have resorted to such an expedient had the economy of our resources to their utmost been a matter of much importance to us at the time. Such, however, was not exactly the case. The sea, against which the guide-rope was especially intended to provide, had long since been passed, and no chance of its recurrence in the least probable. The chief object which we now felt in its continuance was the further trial and proof of the practicability of its employment, which, however, was not so imperative as to prevent us from suspending its action whenever occasion seemed to require it. It must not, however, be thought that these occasions resulted from any deficiency on the part of the guide-rope, or that we should have been *compelled* to discontinue the use of it at any time had particular reasons appeared for adopting an opposite line of proceeding. Where the alternative was a matter of no moment to us, we considered it

in order to maintain our elevation until we had arrived at a place where we could once more conveniently resume the use of the guide-rope. For this purpose, Mr. Green being desirous to employ the lime which had already been used in the receiver of the machine, preparatory to its being charged afresh, and having to that effect opened it over the side of the car, unfortunately let it slip from his hand. Deprived of the most essential part of our apparatus, the lime which was intended to supply it, and of which we had a considerable store, became of no use except for the purposes of ballast, to which account we were subsequently glad to convert it. To dispose of the barrel in which it had been contained was a subject of more serious consideration, its size and weight rendering it rather a dangerous expedient to get rid of it by the ordinary method of discharge. This difficulty, however, we contrived to overcome by attaching it to a small parachute, which served in some degree to moderate its descent, in which guise it was accordingly committed to the earth, where I have no doubt its appearance the following morning within the private precincts of some gentleman's

best to observe that line of conduct which we conceived to be attended with the least possible inconvenience to others, and thus in the present instance avoided coming in collision with a town which shewed even at that late hour of the night such striking symptoms of activity and occupation.

enclosure gave occasion to many a surmise as to the *how* or the *wherefore* of its unexpected arrival.

Having now cleared the town, and once more entered upon the fiery district in which it was embosomed, we again resumed the use of the guide-rope, which, as I have just said, on our approach to so considerable a community we had been temporarily induced to suspend. This operation brought us once more to a nearer contact with the earth, and enabled us clearly to distinguish the voices of many persons whom, notwithstanding the lateness of the hour, we conjectured to be still at work, or else congregated in the neighbourhood of some of the numerous manufactories which illuminated the subjacent country. Desirous to attract their attention, and to enjoy, in idea at least, the surprise with which so novel an apparition was well calculated to inspire them, we lighted and lowered a Bengal light nearly over their heads, at the same time addressing a few words to them through the speaking trumpet, alternately in the French and German languages, one or other of which we thought it most probable they would understand. The effect produced upon them by such an unwonted occurrence was no doubt extreme, as we could readily perceive by the confusion which appeared to reign among them, the hurried tone and elevated expressions which immediately succeeded this unexpected declaration of our presence. *What* they thought of us, however, we had no means exactly to determine; that they were dismayed and perplexed to a considerable degree is neither to be doubted nor wondered at; for in fact such an appearance, and at such a juncture of time, place, and circumstances, might have struck terror into bolder hearts and wiser heads than those of the honest artizans that formed our audience upon this occasion. Catching alone the rays of light that proceeded from the artificial fire-work that was suspended close beneath us, the balloon, the only part of the machine visible to them, presented the aspect of a huge ball of fire, slowly and steadily traversing the sky, at such a distance as to preclude the possibility of its being mistaken for any of the ordinary productions of nature; a suspicion which even if it had existed the terms and tone of our address must speedily have tended to efface. We did not, however, long remain to enjoy their confusion; a consideration of our own convenience more than of theirs, inducing us to give them rather a sudden congé. Amongst the other sounds which issued from this Cyclopean region were some which betokening the presence of a steam engine at work immediately before us, suggested the propriety of raising ourselves to such a height as to place the guide-rope beyond the chance of becoming entangled in some of the machinery. To add therefore to their confusion, while lost in astonishment and drawn together by their mutual fears they stood no doubt looking up to the object of their terrors, a large shower of sand came tumbling down upon their heads, and the tail of the guide-rope at the same moment passing right in the midst of them could not fail to raise their perplexity to the highest pitch. Shortly after, the light expiring, we were lost to their view in the darkness of the sky and the increasing elevation of our ascent. This was the last spectacle of the kind which we were at present destined to enjoy. Scarcely had we passed the confines of the fiery region that had been the scene of our late exploit ere an unbroken obscurity more profound than any we had yet experienced, involved us in its folds, and effectually excluded every terrestrial object from our view.

It was now past midnight, and the world and its inhabitants had finally committed themselves to repose. Every light was extinguished, and every sound hushed into silence; even the cheerful tones of the vigilant watch-dog, which had frequently contributed to enliven our course during the previous portion of the night, had now ceased; and darkness and tranquillity reigned paramount over the whole adjacent surface of the globe.

From this period of our voyage until the dawning of the following day, the record of our adventures becomes tinged with the obscurity of night. The face of nature completely excluded from our view, except when circumstances occasionally brought us into nearer contact with the earth, all our observations during the above period are necessarily confined to a register of incidents and sensations mingled with vague conjectures, and clouded with the mystery wherewith darkness and uncertainty were destined to involve so large a portion of the remainder of our expedition. The moon, to which we might have looked up for companionship and assistance, had she been present, was no where to be seen. The sky, at all times darker when viewed from an elevation than it appears to those inhabiting the lower regions of the earth, seemed almost black with the intensity of night; while by contrast no doubt, and the remotion of intervening vapours, the stars, redoubled in their lustre, shone like sparks of the whitest silver scattered upon the jetty dome around us. Occasionally faint flashes of lightning, proceeding chiefly from the northern hemisphere, would for an instant illuminate the horizon, and after disclosing a transient prospect of the adjacent country, suddenly subside, leaving us involved in more than our original obscurity.

Nothing in fact could exceed the density of night which prevailed during this particular period of the voyage. Not a single object of terrestrial nature could any where be distinguished; an unfathomable abyss of "darkness visible" seemed to encompass us on every side; and as we looked forward into its black obscurity in the direction in which we were proceeding, we could scarcely avoid the impression that we were cleaving our way through an interminable mass of black marble in which we were imbedded, and which, solid a few inches before us, seemed to soften as we approached, in order to admit us still farther within the precincts of its cold and dusky enclosure. Even the lights which at times we lowered from the car, instead of dispelling, only tended to augment the intensity of the surrounding darkness, and as they descended deeper into its frozen bosom, appeared absolutely to melt their way onward by means of the heat which they generated in their course.

Independent of the real obscurity of the night, a combination of two circumstances peculiar to our situation, contributed to occasion the extraordinary impression of darkness which we have here feebly attempted to describe: in the first place, the total absence of all material objects capable of reflecting the scattered rays of light which might happen to exist in the surrounding atmosphere; and secondly, (a natural consequence of the former) the power of availing ourselves of our own light, without dispelling or diminishing the darkness it was otherwise calculated to display. To the former of these were we indebted for the *positive* obscurity of the locality in which we found ourselves; to the second we owe the means of appreciating it by the contrast it enabled us to establish. It is evident that these two circumstances can never be made to exist in combination, except in a situation and under advantages exactly similar to ours: however it might be possible by the most perfect exclusion of light to effect an *artificial* obscurity capable of rivalling that to which we were *naturally* exposed, any attempt to avail oneself of the aid of light to establish the contrast upon which the real strength of the impression depends, must at once subvert the position and nullify the effect it was purposely designed to enhance.

It was now that the advantages of the guide-rope began to make themselves particularly appreciable, in the indications it afforded of the changes that were continually occurring in the level of the subjacent soil, giving us infallible warning of our approach to ground the superior elevation of which might otherwise have occasioned us considerable inconvenience. To such an extent did these alterations at times proceed, that frequently a difference in the altitude of the barometric column would manifest a change of several thousand feet in the level of the balloon's course, while the guide-rope continuing to trail upon the ground, would indicate an uniform distance from its surface of somewhat less than its own extreme dimensions. Several times under the influence of these changes did we arrive* so near the earth, as to be

---

* To prevent misconstruction the reader is requested to observe that the expression here used does not of necessity imply that any change had taken place in the level of the balloon's course to occasion its casual interference with the earth—the changes whereby such a result became possible, being entirely

enabled to distinguish, imperfectly it is true, some of its most prominent features; and as the intensity of the darkness yielded to our approach, obtain some faint idea of the nature of the country which lay beneath as. At these times we appeared to be traversing large tracts of country partially covered with snow, diversified with forests, and intersected occasionally with rivers, of which the Meuse in the earlier part of the night, and the Rhine towards the conclusion, constituted as we afterwards learned, the principal objects both of our admiration and of our conjectures. Nothing could be more interesting than the glimpses which these mysterious approximations would occasionally permit us to enjoy. Slowly descending, as it seemed to us, from a region where darkness formed the only subject of our contemplations, at first some faint hallucination, (but whether of earth or air we could but

attributable to the latter. And yet the phrase is perfectly correct, inasmuch as the *action* by which it was effected was inherent in the former, which in the course of its onward progress became sensible of these changes, and did, strictly speaking, *arrive* in contiguity with the surface of the earth, though without any alteration in the level at which it was proceeding. I have been induced to enter into this explanation from observing that a misconstruction of the kind alluded to has already been put upon the above phrase by the commentator, upon the first edition of this little narrative in one of the daily journals, and an inference drawn therefrom prejudicial to the efficacy of the guide-rope by which these supposed depressions, it is alleged, should have been counteracted.

doubtfully determine), would appear invading the obscurity of the sable vault immediately beneath us, and
giving us the first notification of our approach to something that owned a form and acknowledged the laws of
the material world. Gradually as we drew nigh, these
mysterious appearances would insensibly extend themselves in space, strengthening in their outlines, and
becoming more definite in their form, with an effect, which
to render it more intelligible, we can only compare to that
produced while, looking through a telescope during the
process of its adjustment, the confused and shadowy
features of some distant prospect are made to pass slowly
through every gradation of distinctness ere the proper
focus be at length obtained. Along this indefinite plain,
maintained in our level by the agency of our faithful
regulator, the guide-rope, we would continue to glide for
a considerable time, until some equally unexpected depression in the surface of the earth would gradually abstract
it from our view, and slowly reversing all the impressions
we had before experienced in our approach, once more
consign us to the opaque obscurity that reigned throughout
the upper regions of the air.

An instance of the extraordinary conclusions to which
the vague and indistinct nature of these representations
would occasionally lead us will serve to give some idea of
the doubt and uncertainty that, even at the best, prevailed
over all our observations and conjectures during this most

interesting portion of our voyage. For some time back
our attention had been particularly directed to an appearance, which in the absence of any grounds for suspecting
the contrary, we very naturally concluded to proceed from
some object or other on the surface of the earth below.
Seen through the thick gloom of the night, and extended
alone in the black space that wrapped every other object
from our view, it bore the aspect of a long narrow avenue
of feeble light, starting off in a strait line towards the
horizon, from some point or source at some distance
underneath us. What it could be we fruitlessly endeavoured to determine. For a river, its extreme length
and regularity united, forbid us to assume it; while the
dimensions it must have had to enable it to present so
important an appearance at the elevation we then
occupied, equally precluded the possibility of its being
either a canal or a road, the only other objects to which
we could with any degree of probability refer it. In vain
we looked forward out of the car into the deep intensity
of the surrounding night, concentrating all our powers of
vision to the one spot, that we might catch some clearer
view to determine our conjectures: in vain we racked our
imagination, in the absence of the requisite visual testimony, to devise what it could be that, amid such unbroken
obscurity, contrived to make itself alone distinguishable.
The more we looked, the more we doubted; the more
we reflected, the more uncertain appeared the result of

our speculations; nor was it till after a considerable lapse of time, induced by observing its long-continued presence in the same position, that we became finally aware that it was only one of the stay-ropes* attached to the summit of the balloon, which hanging down along the outside at a distance of five and twenty feet from the car, and being, in fact, the only material object within our ken, had partially caught the rays of light from our lamp, and returned them to us under the aspect and impression we have above endeavoured to describe.†

* Two long cords of moderate dimensions externally attached to the frame of the upper valve, and used to steady the position of the balloon during the inflation, as well as after the descent, during the process of emptying the gas, to prevent her from rolling on the ground. These ropes, when the balloon is full, will extend to some feet below the car, and at a distance of half the diameter of the sphere on either side of the machine.

† If any one will endeavour to imagine himself looking partly forward and partly downward from the summit of a lofty tower, when the obscurity of night is at its highest, and beholding a line partially illuminated, (of the real dimensions of which he is ignorant), vertically suspended at a distance of some yards before him, he will be able to form a pretty correct estimate of the circumstances under which the above erroneous conviction was produced. He will then perceive the impossibility of determining by the mere aid of the senses, the question of the real distance and position of the object, and will be enabled to appreciate the error by means of which the judgment was

In the midst of this intense obscurity an incident occurred, which for the effect it is calculated to produce upon the minds of those who experience it for the first time, and in ignorance of its cause, merits particularly to be noticed.

It was about half-past three in the morning, when the balloon, having gained a sudden accession of power, owing to a discharge of ballast, which had taken place a few minutes before, while navigating too near the earth to be considered perfectly safe in a country with the main features of which we were totally unacquainted, began to rise with considerable rapidity, and ere we had taken the customary means to check her ascent, had already attained an elevation of upwards of twelve thousand feet. At this moment, while all around is impenetrable darkness and stillness most profound, an unusual explosion issues from the machine above, followed instantaneously by a violent rustling of the silk, and all the signs which may be supposed to accompany the bursting of the balloon, in a region where nothing but itself exists to give occasion to such awful and unnatural disturbance. In the same instant, the car,

induced to refer the appearance afforded by a vertical rope a few feet off, *the presence of which it did not anticipate,* to that of some object on the horizontal plane of the earth *which it was constantly expecting to encounter.*

as if suddenly detached from its hold, becomes subjected to a violent concussion, and appears at once to be in the act of sinking with all its contents, into the dark abyss below. A second and a third explosion follow in quick succession, accompanied by a recurrence of the same astounding effects; leaving not a doubt upon the mind of the unconscious voyager of the fate which nothing now appears capable of averting. In a moment after all is tranquil and secure; the balloon has recovered her usual form and stillness, and nothing appears to designate the unnatural agitation to which she has been so lately and unaccountably subjected.

The occurrence of this phenomenon, however strange it may appear, is, nevertheless, susceptible of the simplest resolution, and consists in the tendency to expansion from remotion of pressure which the balloon experiences in rising from a low to a higher position in the atmosphere, and the resistance to this expansion occasioned by the tenacious adhesion of the silk in the folds which the comparatively collapsed state of the balloon had previously allowed it to assume. When the ascent and consequent expansion take place slowly, sufficient time is given to the included gas *gradually* to overcome this resistance, and the balloon is enabled to accommodate itself to the growing dilatation of its contents during the progress of its elevation. When, however, on the other hand, as in the case especially before us, the rapidity of the ascent is such as to anticipate the *gradual* adaptation of the balloon to the expansive tendency of its contents, the entire extrication of the folds of the silk will not take place until the internal pressure of the included gas has reached a considerable amount, when *suddenly* that extrication is accomplished, attended by those effects which we have already attempted to describe. The impression of the descent of the car in the above representation is evidently a false one—on the contrary, elevated by the *longitudinal* curtailment of the balloon in the sudden recovery of its pristine form and *breadth*, the car, so far from sinking, actually springs up; it is the unexpectedness of such a movement, and its apparent inconsistency with the laws of gravitation that occasions the delusion, the reality of which the collateral circumstances essentially tend to confirm.*

---

* In the former editions of this narrative I had attributed the detention of the silk in its corrugated form entirely to the agency of the frost upon the network of the balloon, previously saturated with moisture during its protracted sojourn at a lower elevation. Having, however, since learned from Mr. Green that he has frequently experienced the like effects from a rapid ascent without the intervention of such a cause, I am glad to have the opportunity of generalizing the explanation I had given of the above phenomenon, and of assigning to the frost, in the case alluded to, its proper place as merely contributing to enhance the effect by the additional resistance it offered to the gradual dilatation of the balloon.

The cold, particularly during this part of the night, was undoubtedly intense, as could be perceived not less from the indications of the thermometer, (ranging variously from within a few degrees below, to the point of congelation,) than from the effects which it produced upon the different liquors wherewith we were provided. The water, coffee, and, of course, the oil in our several vessels were completely frozen; and it was only by the actual application of the heat of the lamp that we were enabled to procure a sufficiency of the latter to supply our wants, during the long term of darkness to which we were about to be subjected.

Of the advantages which in these circumstances we had expected to reap from the use of our machine for heating liquors, we had, as I have before observed, been for some time deprived by the loss of a most material part of the apparatus. In this dilemma we had tried several shifts for supplying the deficiency, but unfortunately without effect. Abandoning, therefore, the attempt, we at first became reduced to the disagreeable alternative of drinking our coffee in a state almost approaching to congelation, and finally, as it became more thoroughly frozen, found ourselves compelled to relinquish the use of it altogether.

Strange, however, as it may appear, while all around bore such unequivocal testimony to the severity of the cold, the effects produced upon our persons, undefended as they were by any extraordinary precautions, were by no means commensurate to the cause, nor such as even under ordinary circumstances we might fairly have expected to encounter.

The reason to which may be attributed this unusual exemption from the consequences of a low temperature, is the absence of all current of air, the natural result of our situation, and one of the peculiar characteristics of aerial navigation.

That such a circumstance is fully adequate to the result ascribed to it, ample testimony is afforded in the accounts recently given to the public of the transactions of the great polar navigators, Captains Parry, Back and others, in pursuit of the discovery of the north-western passage, wherein many instances are related of persons under similar circumstances not only bearing, but even enjoying a reduction of temperature many degrees inferior to that in which we were placed. Indeed, from what we are there given to understand, the degree to which the human frame is capable of being refrigerated without experiencing pain or inconvenience, appears to be almost entirely regulated by the concomitant amount of atmospheric motion. In the absence of that motion, there seems to be no limit *in nature* to the extent to which this reduction may be carried; the personages in the above expeditions frequently finding themselves exposed to a

temperature thirty degrees below zero, (or sixty-two below
the freezing point) of Fahrenheit, without even being conscious of anything extraordinary in their situation, until
some change in the state of the surrounding atmosphere
occurred to call it to their senses. After what has been
here stated, I think it will be unnecessary to offer any further disproval of the absurd reports which were circulated
concerning the severity of the temperature to which we
were subjected, and the serious consequences which we
were supposed to have suffered from our exposure to it.

As the night drew on to a close the appearance of the
firmament became subjected to a gradual change. The
stars, insensibly assuming a more natural lustre, began by
slow degrees to " pale their ineffectual fires," while their
light, which, bound as it were by the prevailing obscurity,
had hitherto appeared concentrated and confined, each to
its own particular disk, gradually became more diffuse,
and, illuminating the celestial hemisphere, tended continually to diminish that intense brilliancy which, as we
before observed, had characterised the aspect of the sky
during the crisis of the preceding night. Among these,
the morning star for a long time shone conspicuous, occupying the very centre of our eastern horizon, and creating
around a halo so unwonted as almost to persuade us into
the belief of a premature approach of day. Large
masses of fleecy clouds, now began to be imperfectly distinguished, pervading the lower regions of the atmos-

phere, and for a while leaving us in doubt, whether they were not a continuation of those snowy districts which we so frequently had occasion to remark.

From out of this mass of vapours, more than once during the night our ears had been assailed with sounds bearing so strong a resemblance to the rushing of waters in enormous volumes, or the beating of the waves upon some extensive line of coast, that it required all our powers of reasoning, aided by the certain knowledge we had of the direction we were pursuing, to remove the conviction that we were approaching the precincts of the sea, and transported by the winds, were either thrown back upon the shores of the German ocean, or about to enter upon the remoter limits of the Baltic.

It would be endless to enumerate all the conjectures to which this phenomenon gave rise, or the various manners by which we endeavoured to explain its occurrence. Among them those which seemed to obtain the greatest credit, were that the sounds proceeded from some vast forest agitated by the winds; some rapid river rushing impetuously over a broken and precipitous channel; or finally, that the misty vapours themselves, by the mutual action of their watery particles, or their precipitated deposition upon the irregular surface of the earth beneath, had occasioned the murmurs, which, multiplied throughout so large a space, came to our ears in the formidable accents to which we have above alluded.

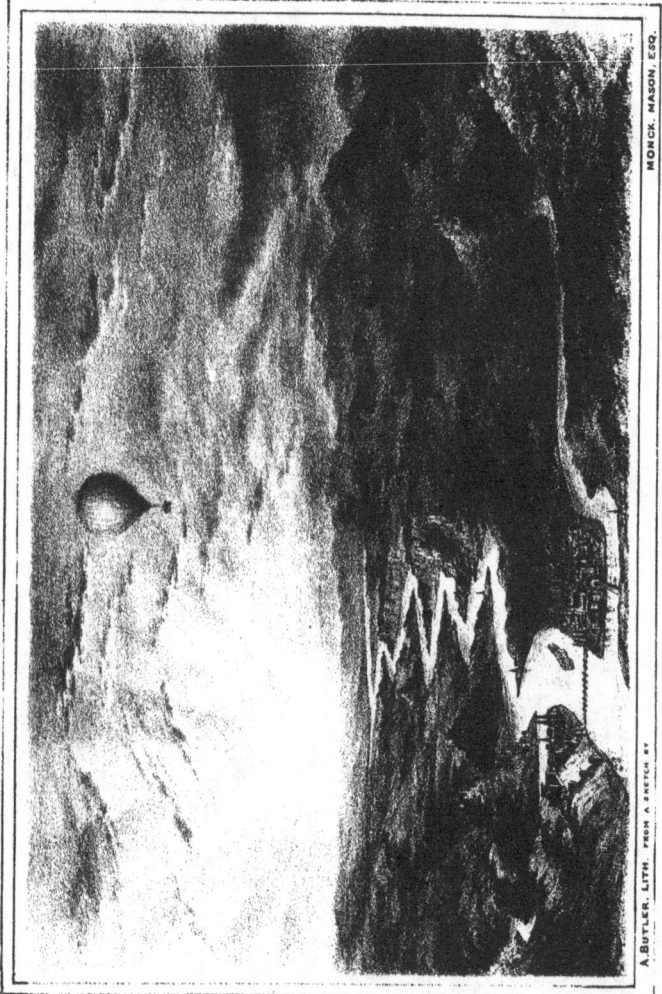

A. BUTLER, LITH. FROM A SKETCH BY   MONCK. MASON, ESQ.

THE BALLOON PASSING OVER COBLENZ.

According as the day drew nigh these appearances vanished, with much of the doubts to which they had given rise. Instead of the unbroken outline of the sea, an irregular surface of cultivated country began feebly to display itself, in the midst of which the majestic river we had noticed for some time back, appeared dividing the prospect, and losing itself in opposite directions amid the vapours that still clung to the summits of the hills, or settled in the valleys that lay between them. Across this river we now directed our course, and shortly after lost sight of it entirely, behind the gently swelling eminences by which it was bordered on both sides.*

The dawn, which for some time back had been continually augmenting, had now become fully established in the upper regions of the atmosphere, although its influence as yet was but slightly exerted upon the humbler districts of the subjacent earth. All the celestial bodies

* In the grey dawn of the morning, shortly after we passed this river, we were noticed, as we afterwards learned, by two sportsmen who happened at that hour to be out exercising their calling on the neighbouring hills. Not knowing what to think of the apparition, and deceived by his ignorance of its size into a belief that it was not very distant, one of them was inclined to shoot at us, and for that purpose had just levelled his gun, when his companion observing some of our movements arrested his aim, and after a short reflection convinced him of the real nature of the being he would have had to contend with if he had had the misfortune to bring us about his ears.

had now entirely disappeared ; even the morning star, which so long the subject of our admiration, had continued with waning energy to contest the empire of the sky, had now retired, and we began earnestly to look forward to the arrival of the great luminary that was soon to supply their place.

About ten minutes past five one of those casual aberrations occurred, to which we have already alluded, when the balloon rising rapidly, we became suddenly transported to an elevation of above twelve thousand feet. This was the highest point we attained throughout the whole voyage, and the effect was in truth equally preeminent with the occasion by which it was produced. If we only reflect that our position at this altitude was such as to have enabled us to behold objects at a distance of above one hundred and fifty miles on every side of us, had those objects been sufficiently great or sufficiently striking to fix the attention, some faint idea may be had of the immensity of prospect which at that moment became subjected to our view. We shall then be seen occupying the centre of a circle, whose diameter extending to above three hundred miles in length, afforded us an horizon, the circumference of which exceeding an equal number of leagues, comprised within its circuit an expanse of visible surface little short of seventy-one thousand square miles. In the enjoyment of this stupendous landscape we continued for above an hour, occasionally descending a few hundred feet, and again rising to resume our station upon our former level.

In one of these latter movements, which took place at about a quarter past six,* the balloon having nearly recovered its highest elevation suddenly brought us in full view of the sun, and for the first time gladdened with the assurance of a speedy return of day.

Powerful, indeed, must be the pen which could hope to do justice to a scene like that which here presented itself to our view. The enormous extent of the prospect—the boundless variety it embraced—the unequalled grandeur of the objects it displayed—the singular novelty of the manner under which they were beheld—and the striking contrast they afforded to that situation and those scenes to which we had so long and so lately been confined, are effects and circumstances which no description is capable of representing in the light in which they ought to be placed, in order to be duly appreciated. Better far to leave it to

---

* The time referred to here and elsewhere throughout this narrative, is that of Greenwich.—Upon the completion of the voyage, a variation, amounting to about thirty-four minutes, was found to exist between the times indicated at its two extremes; the chronometers of Weilburg, being so much in advance of those of London. This variation was occasioned by the easterly direction of our course, and the difference in latitude, to the extent of eight degrees, twenty minutes, between the two places,

a fertile imagination to fill in the faint outlines of a rough and unfinished sketch, than by a lame and imperfect colouring, run the risk of marring a prospect, which, for grandeur and magnificence has certainly no parallel in all the vast and inexhaustible treasures of nature.

This splendid spectacle, however, we were not long destined to enjoy; a rapid descent, which shortly after ensued, for a while concealing it from our view, and once more consigning us to the shades of night, which still continued to reign unbroken throughout the lower region of the air.

Again we rose within the reach of this delightful prospect; and again did we lose sight of it, amid the vapours and obscurity that accompanied our descent; nor was it till we had three times made the sun rise, and twice beheld it set, that we could fairly consider it established above the horizon, and daylight complete upon the plane of the earth beneath us.

From this time forward all our observation was principally directed to the nature of the country, and its adaptation to the descent which we had now resolved to effect, the first fitting opportunity. To this step, the uncertainty in which we necessarily were, with respect to the exact position we occupied, owing to our ignorance of the *distance* we had come, especially determined us. For a long time past, the appearance of the country, so unlike any with which we were acquainted, had led us to entertain serious doubts as to whether we had not already passed the limits of that part of Europe where we might expect to find the accommodation and conveniencies which our own comfort, and the safety of the balloon, imperatively demanded. This opinion, the large tracts of snow over which we had passed, during the latter part of the night, bearing a strong resemblance to all we had hitherto pictured to ourselves of the boundless plains of Poland, or the barren and inhospitable Steppes of Russia, considerably tended to confirm;* and as the region we were immediately approaching seemed to offer advantages which, under these circumstances, we could not always hope to command, we resolved not to lose the occasion it so opportunely appeared to have afforded us.

As soon as we had come to this determination all preparations were speedily commenced for the descent; the guide-rope was hauled in, (an operation of much labour, owing to the bad construction and imperfect action of the

---

* This presumption will not appear so extravagant when we consider the enormous rapidity with which the course of the balloon is liable to be affected, and the impossibility of obtaining any indication as to its amount during the long period of darkness which we had just encountered. Had we continued to pursue the greatest rate of motion at which the balloon has been known to be impelled in these latitudes, we should ere the period of our descent have accomplished a distance of above two thousand miles.

windlass;) the grapnel and cable lowered, and every thing got ready that we might be able to avail ourselves of the first and fittest opportunity that might occur. To this intent, likewise, we quitted our exalted station, and sought a more humble and appropriate level, along which we continued to range for some time, and to a considerable distance; the yet early hour of the day deterring us from completing the descent, in the fear of not obtaining that ready assistance from the inhabitants which it is always the main object of the aeronaut, if possible, to secure.

As the mists of the night began to clear away from the surface of the soil, we were delighted to perceive a country intersected with roads, dotted with villages, and enlivened with all the signs of an abundant and industrious population. The snowy covering which so lately chilled us with its forbidding aspect, had now disappeared, except a few patches which still lingered in the crevices, or lay spread within the sheltered recesses of the numerous hills by which the surrounding neighbourhood was particularly distinguished. On the summit of one of these, an isolated edifice of considerable magnitude and venerable antiquity appeared just breaking through the vapours that yet partially concealed the morning landscape. Seated upon the very point of the eminence, it seemed like some ancient baronial castle overlooking the prospect, and extending its protection to a cluster of humbler dwellings

that straggled around its base. One or two towns, likewise, of superior pretensions, were distinctly to be seen;
giving promise of accommodation and advantages, which
in our present emergencies, and under our present convictions, were not to be neglected. Accordingly, having
pitched upon the spot most proper for the purpose, the
valve was opened, and we commenced our descent.

The place so selected was a small grassy vale, of
about a quarter of a mile in breadth, embosomed in hills,
whose sides and summits were completely enveloped
with trees. Beyond this, on the opposite side, lay
another valley of the same description; the only one
visible for many miles, where we could conveniently effect
our landing; an endless succession of forest scenery completing the landscape in the direction in which we should
have had to proceed. Into the former of these we now
precipitated our descent, with the design of alighting, if
possible, in the centre, clear of the woods that enclosed it
on all sides. In these hopes, we were, however, disappointed : the wind suddenly increasing as we approached
the ground, so much accelerated the onward course of
the balloon, that before the grapnel could take effectual
hold of the soil, we had passed the middle of the valley,
and sweeping rapidly over the ground, were borne close
against the wooded declivity that flanked its eastern termination. To discharge a sufficiency of ballast to raise
the balloon, and carry her clear of the impending danger,

was the natural remedy. An unexpected obstacle to this operation here again presented itself: the sand which forms the ballast, frozen during the night into a solid block of stone, refused to quit the bag in the proportion required, and no time remained to search for one more suited to the occasion. Not a moment was, in fact, to be lost; the valley was passed, and the branches of the trees that clothed the opposing precipice, were already within a few feet of the balloon; the grapnel continued to drag, and no chance appeared of arresting her progress onward. In this emergency one alternative alone remained, and the sack itself, with all its contents, to the amount of fifty-six pounds in weight, were at once consigned to the earth. In a moment, the balloon, lightened of so large a portion of her burden, had sprung up above a thousand feet, and clearing the mountain at a bound, was soon in rapid progress to the realms above. To counteract the consequence of this sudden accession of power, and avoid being carried beyond the reach of the second valley, which we have already described as the only other available spot for our descent, the valve was again opened, and issue given to a large quantity of gas; sufficient, as was calculated, to check the course of the balloon in time to enable us to attain the point to which all our views were now directed.

A second time, however, we were doomed to be disappointed. No sooner had we completed this manœuvre, than by another caprice of nature, the wind suddenly abating, we found ourselves at once becalmed, and rapidly descending into the bosom of the woods that capped the summit, and clothed the sides of the intervening eminences. From this dilemma we were only relieved by the timely discharge of a further portion of our weight; not, how- ever, before the accelerated descent of the balloon had brought us within a cable's length of the ground,* and almost in contact with the upper surface of the wood. Here, for a few moments, we continued to hover; the grapnel struggling with the top-most branches of the trees, and grasping and relinquishing its hold according to the varying impulse of the slight wind that prevailed at our elevation.

While in this situation, we perceived, standing in a path in the wood, two females, the first inhabitants we had noticed, lost in astonishment, and seemingly petrified with gazing upon so astounding an apparition. It was in vain we addressed them with a speaking- trumpet, in the hopes of procuring the assistance of some of the male population, which we conjectured could not be far off; the sound of our voices, proceeding from such an altitude, and invested with such an unearthly character, only augmented their astonishment, and added

* The length of the cable to which the grapnel is attached is about one hundred and twenty feet.

to their fears; they fled incontinently, and without waiting farther parley, sought the shelter of the neighbouring coverts.

After continuing for a few minutes longer in these straits, we at length reached the confines of the wood; when, resolving not to be again baffled in our designs by the treacherous inconstancy of the wind, the valve was opened to its fullest dimensions, and the grapnel taking hold shortly after, we came to the ground with considerable, though by no means, disagreeable rapidity.*

Too much praise cannot be given to Mr. Green, for his excellent conduct throughout the whole of this intricate pilotage. It is not by reading a mere description of the difficulties encountered, and the manner by which they were counteracted, that a correct judgement can be formed upon the merits of such a case as this; a further consideration is necessary—the knowledge that these difficulties did not proceed from the same source as the remedies by which they were defeated. In this light it is, that the conduct of our celebrated captain, has a right to be criticised; the impediments were those of uncontrollable nature—the victory, and the means employed to secure it, were all his own.

* It was half-past seven o'clock when this occurrence took place, and our descent could be fairly said to be completed. The duration of our voyage may therefore be calculated at exactly eighteen hours.

As soon as the descent was completed, and the power of the balloon sufficiently crippled to permit one of the party to quit the car, the inhabitants, who had hitherto stood aloof, regarding our manœuvres from behind the trees, began to flock in from all quarters; eyeing, at first, our movements with considerable suspicion, and not seldom looking up in the direction from which we had just alighted, in the expectation, no doubt, of witnessing a repetition of this, to them, inexplicable phenomenon.

A few words in German, however, served to dissipate their fears, and secure their services. The first question, "Where are we?" was speedily answered, "In the Duchy of Nassau, about two leagues from the town of Weilburg." The second was theirs, "Where do you come from?" "From London, which we left yesterday evening." Their astonishment at this declaration may be easily conceived. The fact, however, was not to be disputed. What they had seen was to the full as marvellous as any thing we might choose to relate, and certainly enough to entitle us to consideration, and command respect.

At all events, whether from *above* or *below*, we were evidently strangers; a circumstance of itself sufficient at all times to have engaged the sympathy and assistance of an artless and hospitable people, but which, coupled as it was in our case with the possibility of one or other of the two preceding alternatives, brought us in for no small amount of homely deference and attention.

To these kindly feelings we endeavoured to contribute by every means in our power. Our stock of biscuits, wine, and brandy, quickly disappeared; with a relish which the novelty of the journey they had so lately performed, tended, no doubt, considerably to augment. The brandy, in particular, so much stronger than any they had ever before essayed, attracted their special admiration; and as they, each in succession, drank off their allowance, they seemed by the exclamation of "Himmlischer Schnapps,"* which accompanied every draught, as well as by the upward directions of their eyes, to denote the quarter from which they now became fully convinced, a beverage so delicious could alone have proceeded.

With all the willingness, however, which they displayed in their endeavours to assist us, it required no little management, and a thorough knowledge of the peculiar habits and propensities of the *animal*, to turn their services to a proper account. In the first place, the operation of emptying the balloon, at all times sufficiently tedious, was rendered more so in the present instance from the quantity of frozen moisture it had imbibed in the course of the night, and which we were desirous to get

* The literal interpretation of the above expression is—" Celestial dram."

A. BUTLER, DEL.ᵗ & LITHOG. FROM A SKETCH BY
MONCK, MASON ESQ.

DESCENT OF THE BALLOON IN THE VALLEY OF ELBERN.

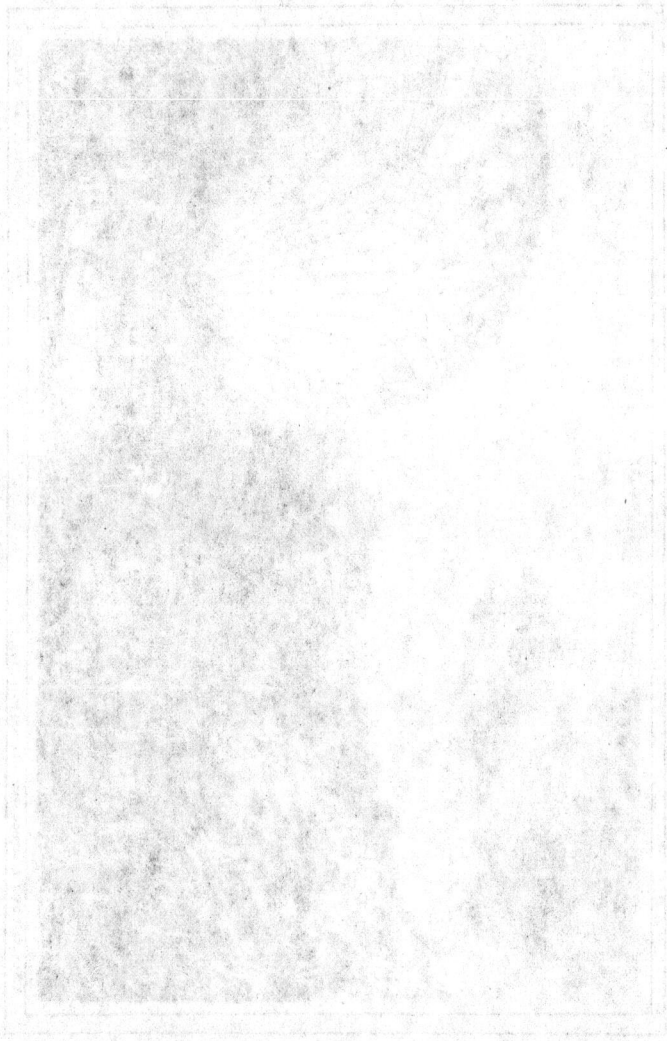

rid of by a little exposure to the sun before we had completely enclosed it. Now Germans, proverbially indolent, require no small degree of excitement to keep their attention and their services engaged to any continued pursuit. The slightest relaxation, therefore, on our part, was sure to be attended with a corresponding relaxation upon theirs, and in the event of our taxing their patience too severely there was no small probability that they would slacken in their efforts, and getting tired of seeing nothing done, eventually abandon us to our resources. On the other hand, to occupy their attention by a liberal distribution of "Schnapps," (the only alternative that remained to us), was not without its particular inconvenience. Germans are never without pipes in their pockets, and never think of eating or drinking without concluding the operation by abstracting the said pipes and indulging in a friendly fumigation; in which case, besides incurring the risk of combustion from so many fire-works in exercise at once, we should have had to calculate upon the certain loss of one hand to each individual, and the other deprived of half its energy, when two, well applied, were scarcely enough for the purpose. Into this error we had fallen at first; the consequence was that half of our efficient forces were already laid up smoking, and it was only by a timely withdrawal of the supplies that we were enabled to command the services of the remainder.

With all these drawbacks, it was nearly twelve o'clock before the whole of our operations were concluded and the balloon, with all its accompanying apparatus, safely adjusted in the bottom of the car. Our next step was to procure a cart and horses to convey it to Weilburg, the nearest place where we could expect to meet with the accommodation which the circumstances of the case rendered desirable. For this, as there was but one in the neighbourhood for many miles around suitable to the purpose, we were compelled to submit to a further delay of about an hour and a half. In the meantime we had some difficulty in inducing our kind and able coadjutors to accept of any remuneration for the timely assistance they had afforded us; nor was it until we had evinced by our perseverance a determination not to be refused, that we finally succeeded in persuading them to come to some definite arrangement among themselves as to what amount of compensation should be bestowed, and in what manner it should be distributed. Accordingly, as the magnitude of their numbers precluded the possibility of extending our bounty to all, fourteen were selected out of those who had taken a most prominent part on the occasion, and the sum of half a franc each, equal to about five pence English, stated as the full amount of their expectations. This sum, with more liberality than prudence, as appeared in the sequel, was immediately doubled,

when a scene occurred to which no description is capable
of communicating the entire effect. Scarcely had this
unexpected extension of our bounty been announced to
the fourteen fortunate individuals who were to participate in it, ere as many unwashed beards, black and
brown, white, yellow, red, and grey, were simultaneously and unceremoniously thrust forward for the purpose of signifying their gratitude by effecting a salute, in
a style which, in our country at least, is usually considered
one of the peculiar privileges of the gentler sex. To
refuse the proffered courtesy might have been construed
into an affront, and we were absolutely in the very act of
being subjected to this agreeable ordeal, when the season-
able arrival of the long expected vehicle saved us from the
full infliction of the direful penalty. Overjoyed at our
timely deliverance, all hands were summoned to assist in
loading the waggon, and having mounted thereon ourselves, we quitted this, to us, ever memorable spot,* and
attended by an amazing concourse of persons of every
rank, age, and sex, set out for Weilburg, which a few
hours enabled us to attain.

* The exact spot where the event took place was in a field
adjacent to a mill, known by the name of Dillhausen, situated
in the valley of Elbern, in the commune of Niedershausen, about
two leagues from the town of Weilburg; already, by a curious
coincidence, noted in the annals of aerostation as the place
where the celebrated M. Blanchard effected his landing after
an ascent which he made at Francfort in the year 1785.

The fame of our adventure had, however, already preceded us. On our approach we found ourselves greeted
with acclamations, and a ready welcome and honourable
attentions awaited our arrival. All the resources of the
town were immediately placed at our disposal; the use of
the archducal manège was tendered for the occupation of
the balloon; and sentries, more indeed as a guard of
honour than of protection, stationed at the doors and
avenues leading to the place of its reception.

Here then we resolved to remain until our future
movements should be determined by the return of the
letters we had dispatched to Paris immediately upon our
descent. In the mean time, favoured by the peculiar
advantages of the building, we availed ourselves of this
delay to open and inflate the balloon, as well for the
purpose of drying and examining it, as to make some
return for the obligations we were under, by contributing
to gratify the curiosity of our hospitable entertainers.
It would be scarcely credible were I to relate the interest
wherewith the inhabitants seemed to regard this, to them,
novel exhibition; the numbers that poured in to witness
it from all quarters, for many a league around, or the
grateful acknowledgments with which they never ceased
to overwhelm us during the fortnight it continued open
to public inspection.

Nothing in fact could surpass the courtesy and attention
that we experienced from this simple-hearted and hospi-

table community, during the whole period of our residence
at Weilburg. Every one seemed to vie with each other
in conferring favour, and contributing to our entertainment. Balls, dinners, concerts, and other amusements,
were given without intermission, poems were composed
in honour of our adventure,* and the congratulations of
the city presented to us by a deputation of the principal
citizens, headed by their chief civil officer, in the form of
a document duly signed and sealed by the competent
authorities.

Among the festive recreations to which our unexpected
arrival at Weilburg gave rise, we must not omit to mention the ceremony of christening the balloon, which took
place the day previous to our departure: The Baron de
Bibra, Grand Maître des Eaux et Forêts, and the
Colonel, Baron de Preen, being the godfathers; the
Baroness de Bibra and the Baroness de Dungern, the
godmothers, on the occasion. The balloon having been distended with air to the greatest size the dimensions of the
place would admit, eight young ladies, in company with
Mr. Green, entered within the gigantic sphere, and the
name of " The Great Balloon of Nassau" having been
bestowed by one of their number, Mdlle. Theresa, the

---

* Some of these effusions which appeared most deserving of
notice have been subjoined at the end of the volume. (*see*
Appendix G.)

lovely and amiable daughter of the Baron de Bibra, accompanied by a copious libation of wine, the ceremony was concluded with a collation, consisting of the remains of our stock of provisions, which had been unconsumed at the time of our descent.

One other act of honourable attention yet remains to be recorded. On the evening of the same day, the last we had to enjoy in the society of this courteous and hospitable population, it was resolved to signalize the occasion of our visit, and the agreeable intercourse to which it had given rise, by some more flattering display of favour than any we had yet experienced. A grand festival was consequently held in the principal chambers of the chief inn, which had been tastefully decorated for the purpose, and at which all the first personages of the town were assembled to meet us. After the dinner, or rather the supper had been concluded, and the mutual goodwill of the parties established by a general interchange of glasses, occasion was taken to pronounce a short discourse in Latin verse,* composed by M. Friedemann, Principal of the academic Gymnasium, in which a comparison is instituted between our late enterprise and others of a similar nature; at the conclusion of which a crown of laurels was placed upon the head of Mr. Green, and his health with that of his companions proposed and drank amid general and repeated acclamations.

* See Appendix G. No. 1.

From such an universal display of hospitality and kindness it would be difficult to single out any to whom in particular our thanks are due: among those, however, whose station and circumstances entitle them to especial notice, were the Baron de Bibra, Grand Maître des Eaux et Forêts; the Baron de Dungern, Grand Ecuyer de son Altesse, pensionné; the Colonel Baron de Preen, and their respective ladies; M. Hutschsteiner, Premier Conseiller de Médecin; M. Giesse, Premier Conseiller de Justice; M. Friedemann, Superior of the University, and M. Barbieux, likewise attached to the same establishment, together with a variety of others, the mere repetition of whose names would prove but a little recompense for the kindness we received at their hands.

Through the Baron de Bibra, likewise, we took the opportunity to present to His Highness the Duke of Nassau the flags * which accompanied the expedition, as a slight token of the hospitable reception we had expe-

* Besides the usual national insignia, these flags displayed a series of allegorical representations descriptive of the rise and progress of aerostation. Independent, however, of any merit which they might possess from their execution or design, there was one circumstance in their history which rendered them invaluable in the eyes of the aeronaut: they had already performed two hundred and twenty-one voyages in the air; having been the constant companions of Mr. Green's excursions ever since his fifth ascent, wherein he had the misfortune to lose his balloon and all it contained in the sea off Beechy Head.

rienced in his territories, with a request that they should be preserved in commemoration of the occurrence, among the archives of the Ducal Palace at Weilburg, where they now lie alongside of that which half a century before M. Blanchard deposited in like manner, to perpetuate the remembrance of a like event.

On the following morning, November 20th, at an early hour, we took leave of Weilburg and its hospitable inhabitants, and set out for Coblentz on our way to Paris, whither it was now determined we should proceed. At Coblentz, where we arrived late the same evening, it was our intention to purchase a carriage and having stripped it of its body, place the car containing the balloon and other accessories upon the springs, and in that guise, availing ourselves of the same conveyance, continue our journey by post.

This, with some difficulty and the delay of a couple of days, we at length accomplished, and by a proper adjustment of the contents, fixing a temporary seat athwart, and protecting the whole with a covering of oil-cloth, constructed as we considered, (how correctly will appear in the sequel,) a very convenient retreat for the accommodation of such of the party as should be destined to enjoy it. Here also we parted from our companion Mr. Hollond, whose business requiring his immediate return to England, I gladly undertook to accompany the balloon to Paris.

Accordingly, all our arrangements being at length

completed, early on the morning of the 24th, Mr. Green and myself again set out, intending to continue our journey night and day until we arrived at its conclusion. A series of misfortunes, however, appears to have awaited us from the very outset. The weather which had hitherto been particularly fine, considering the time of year, suddenly and completely broke up, and torrents of rain accompanied by powerful and piercing winds ushered in the morning of our departure. Its effects were soon but too perceptible upon our hastily constructed equipage. In the first place our water-proof covering, but ill deserving the name, turned out no better than it should be; in short, any thing but water-proof. Prepared merely with common size instead of varnish, the first half-hour's exposure to the rain completely divested it of every particle of dressing, and discovering a mere groundwork of canvass nearly as porous as netting, left us almost wholly unprotected to " abide the pelting of the pitiless storm." Gusts of wind at every step likewise poured in from all quarters, shaking our frail tenement to its inmost fastenings, and threatening every minute to deprive us of the nominal protection of the little covering which the rain had left us. To complete the comforts of our situation, scarcely had we quitted the town ere our seat, which had been too slightly constructed for the roads we were about to encounter, suddenly gave way beneath us, precipitately consigning us to the bottom of the car, where we lay for some time perdue among the various articles with which that part of the conveyance was plentifully bestrewed.

As it was vain to think of trying to remedy these disasters in the country where we then were, our only alternative was to push forward as fast as we could, until we should arrive at some place where we might obtain materials to repair our shattered vehicle. It was not, however, till the conclusion of the second, or rather the morning of the third day, that we were able to accomplish this. At the village of Thionville, where we had been forced by the inclemency of the weather to pass the preceding night, we at length procured a quantity of common striped holland, the only stuff we could find suited to the purpose. With this we completely covered in the whole machine, and having caused it to be stitched down on all sides, except a small opening in front, whereby to creep in and out, extended ourselves at full length upon some clean straw, which served to separate us from the balloon and other articles beneath, and in that condition prosecuted our journey; to the no small delight and astonishment of all the little boys and girls that, at every stage we came to, and every village we passed, flocked in numbers to greet us; much edified no doubt by the spectacle we afforded them, though sadly at a loss to comprehend how a basket so heavy that four horses were scarcely sufficient to draw it, should have been able to convey us through the air to such a distance. In this manner, " Heu! quantum mutatus ab illo," we continued our route to the French metropolis, sorrowfully contrasting our present with our late conveyance, and indulging in many a comparison between the comforts of aerial and terrestrial travelling, much, it must be confessed, to the advantage of the former.

After journeying in this way for six long days and longer nights, we at length reached Paris, where new honours and a hospitable entertainment awaited our arrival.*

———

Thus ended an expedition which, whether we regard the extent of country it passed over, the time wherein it was performed, or the result of the experiment for the sake of which it was undertaken, may fairly claim to be considered among the most interesting and important which have hitherto proceeded from the same source. Starting from London, and traversing the sea, which mere accident alone prevented from forming a more im-

* Among the other testimonies of honourable distinction which the various scientific and other bodies in that city conferred upon us in respect of our undertaking, I must not forget to mention the medal which was bestowed upon Mr. Green by the Society of the " Académie de l'Industrie Française," for his ingenious discovery of the guide-rope, with the principles of which they expressed themselves perfectly satisfied.

portant feature in our route, in the short space of eighteen hours we performed a voyage which, including only those deviations we have since been enabled to ascertain, rather exceeds than falls short of an extent of five hundred British miles.

It would be endless as well as useless, to enumerate all the places of name or notoriety, which a subsequent examination of the map, aided by the reports of our appearance at different stations by the way, shewed us to have either passed over or approached at some period or other during this extraordinary peregrination. A considerable portion of five kingdoms, England, France, Belgium, Prussian Germany, and the Duchy of Nassau; a long succession of cities, including London, Rochester, Canterbury, Dover, Calais, Cassel, Ypres, Courtray, Lille, Oudenarde, Tournay, Ath, Brussels, with the renowned fields of Waterloo and Jemmapes, Namur, Liège, Spa, Malmédy, Coblentz, and a whole host of intermediary villages of minor note, were all brought within the compass of an horizon, which our superior elevation and the various aberrations we experienced, enabled us to extend far beyond what might be expected from a mere consideration of the line connecting the two extremities of our route.

To all this there was but one drawback, in the time of year in which the experiment was conducted, and which, by curtailing our daylight, devoted to the obscurity of night so large and interesting a portion of the expedition. Over this, however, we had no controul; the constant occupation of the balloon for the purposes of public exhibition during the summer months, left no chance of its being procurable at a better season of the year, especially for a project such as ours, the determination of which as to time and distance was a matter of complete uncertainty. The excursion must therefore have been undertaken as it was, or altogether abandoned; of these alternatives Mr. Hollond unhesitatingly preferred the former.

Ere concluding this hasty narrative, a word or two is required concerning the success of that experiment which formed the main feature, as well as the chief object of the expedition. That object I have already stated to have been the verification, by proper trial, of the power of the guide-rope in determining the course of the balloon within certain restrictions, and the feasibility of its employment under every aspect of circumstances, to such an extent as to render it a valuable and efficient instrument in the hands of the practical aeronaut. In both these respects I have no hesitation in declaring the success of the experiment to have been complete, and the discovery itself one, the entire result of which, on the future progress of the art, it would be impossible at present to anticipate. With such an instrument as this, there now seems to be no limit to the powers of aerostation; no

H

bounds to its sphere of action. All the theoretical objections which a hasty consideration of the means might
otherwise have suggested, experiment has already proved
to be erroneous; and, perhaps, the best illustration that
can be afforded of the powerful influence which this discovery is capable of exerting in favour of the art is, that
under its auspices and with all other advantages to the
extent we enjoyed them on the late occasion, I should
not feel the slightest diffidence in committing myself to
the conduct of the winds, with the intention of continuing
my voyage until I had completed in my course the circuit
of the world itself.

APPENDIX A.

THE conveyance through the atmosphere by means of the
balloon is a thing so entirely "sui generis," so essentially
distinct in all its bearings from every other process with
which we are acquainted, that no force of reasoning is of
itself capable of awakening in the mind of an utter stran-
ger to the art, any adequate notion of the peculiar phenomena which characterize this novel and interesting mode
of transport. So devoid indeed may it be said to be of any
of those analogies which in other matters serve to supply
the place of actual experiment in determining the general
results of new and untried combinations, that I am convinced if an individual were to set himself down with the
intention of endeavouring to picture in his imagination the
various circumstances and impressions which develop
themselves in the practice of aerostation, with all the
advantages which a thorough knowledge of the arts and
sciences in general could contribute to his assistance, he
would still arrive but at a very rough and imperfect representation of the real nature of the case in question. With
so few opportunities of forming a more correct estimate
by personal experience, it is not to be wondered at should

much ignorance be found to prevail upon this head, even amongst those who seem at least to take the strongest interest in its details.

Much of this obscurity, it is true, might have been removed, and the mysteries of the art brought within the reach of ordinary inquirers, had the experience of others been but turned to its proper account, and rendered available to the purposes of general information. Unluckily, however, both for the interests of aerostation and the credit of its votaries, this has not been the case, either to the extent or in the manner likely to prove beneficial to the cause for which it is required. Indeed so far from having contributed to elucidate, I am sorry to say that it is to the conduct of its professors in general that may be traced the greater part of those erroneous impressions that have so long continued to influence the public opinion in regard of all matters appertaining to the subject of our present inquiry. Following a pursuit which, however at present its frequency and our own more enlightened notions may have enabled us to appreciate more justly, in the earlier stages of its practice must have almost appeared to border upon the marvellous, and indebted for their gains to the celebrity which the wonders they were supposed to have witnessed and the dangers they were reputed to have incurred, were in their opinion the readiest means of obtaining, a more natural than creditable propensity to amplification appears to have guided the majority even of those whose adventures without such exaggeration would have entitled them to a fair share of worldly honours and of worldly remuneration. The spirit of emulation coming in aid of incentives already, of themselves, sufficiently powerful, there are really no bounds to the extravagance and absurdity of the tales which have been put in circulation by the different members of this high-flying community, in illustration or defence, each of their own particular exploits.*

But in truth, for all kinds of imposition, the present, above all other pursuits, affords the most especial facilities. Exercised almost invariably alone, or if in company, with individuals whose tastes and interests alike would rather lead them to favour than impugn the credit of whatever it might be thought desirable to allege, and carried on in realms of space, far beyond the reach of human observation, where not an eye exists to witness, or an ear to overhear them, the opportunities for the institution of error are no less boundless than the means of conviction are limited and incomplete. Even the very region to which the operations of the aeronaut are devoted con-

---

* From this charge I feel it a duty, most unequivocally to exempt Mr. Green, who, I am bound to confess, has at all times testified the strongest antipathy to all species of fabrication, and indeed has ever shewn himself much more inclined to undervalue than exaggerate the peculiarities of the profession he has so long and so ably supported.

duces towards the same conclusion, and, rich in grounds
for the marvellous, tends no less to encourage the development of the passion than to administer to its support,
facilitate its exercise, and shelter its mal-practices.

With inducements, therefore, equal to their opportunities, it is not surprising that the cultivators of aerostation
as a *profession* should frequently have availed themselves
of the peculiarities of their situation, and occasionally
transgressed the bounds of strict veracity in the endeavour to raise themselves in the estimation of an over-
credulous and enthusiastic public. Some of the fictions
which have proceeded from this source are mere exaggerations of sensations and effects, either observable in
the ordinary course of aerial navigation, or which the adventurers conceived would be the results of certain situations, and in order to induce the belief of their having
been in those situations, feigned themselves to have experienced. Of this nature are the many tales of wonderful
impressions produced, and extraordinary sufferings endured in the attempt to penetrate into the mystic regions
of upper air to an extent beyond the ordinary course of
aeronautical adventure. Some of these are ludicrous
enough, and display in their construction no bad example
of the " danger of possessing a little science."

One gentleman, for instance, whose name out of respect
to the survivors of a respectable family we hesitate to
repeat, asserts that upon one occasion, in company with

his son, he ascended to such an elevation that the wrinkles of his face had entirely disappeared through the preternatural distension of his skin, and to the astonishment of his companion, although long past his grand climacteric, he had actually begun to assume the aspect and appearance of his former youth !

Another of these aspiring heroes, influenced no doubt, by the same motives, and unwilling to be outdone in the race of aerostatic notoriety, positively declares that he had risen to such a height that his head, swollen by the diminished pressure, had become so much *enlarged*, that he could no longer wear his hat; while a third, out-heroding Herod, and soaring in imagination as in the realms of space, actually rises so high that he is unable to keep his hat from tumbling over his eyes, so much has his head become *contracted* by the extraordinary consequences of so unusual an elevation ! All these and others of a similar nature, are as I have before said, either exaggerations of existing or anticipations of probable results, and owe their origin to imperfect analogies or inconclusive reasoning. In the former instance, for example, the gentleman so marvellously regenerated had no doubt witnessed the experiment of the restoration of a withered pear beneath the exhausted receiver of a pneumatic machine, and without sufficient investigation hastily concluded that were he placed in circumstances of similar rarefaction he would of course exhibit similar phenomena. An essential ingredient in the fact he had, however, overlooked, namely, the impervious nature of the rind of the pear. He was not aware, or perhaps did not reflect, that the expansion, by means of which the wrinkles in the fruit were made to vanish, was not owing merely to the diminution of atmospheric pressure "per se," but to the resistance afforded to the escape of the included gas by the impermeability of the receptacle in which it was inclosed. For the gentleman to have rendered himself liable to the same result, it is absolutely necessary that he should not only have varnished the skin of his face, but even that of all the rest of his body besides. In the same manner as the preceding, the remaining examples we have above quoted, owe their origin to mistaken analogies between vulgar experiments in the receiver of an air-pump, and the supposed consequences of an exposure to a similar diminution of pressure in the open air; the fallacy of the conclusion will be exposed further on, when we come to treat of the real effects of atmospheric rarefaction.

In the construction of these fables, a certain degree of allowance may be made for men, who having really ascended to considerable elevations without having experienced results equal to their expectations, and not possessing science enough to account for their absence, thought that they would not be credited by the world were they simply to state the real circumstances of the case, and accordingly felt themselves compelled in self-defence to feign the occurrence of all those sensations and results which they expected, and believed the rest of the world would likewise expect them, to have encountered had they really arrived at the elevation they either had, (or would have the credit of having) attained. To a certain extent, therefore, an error of judgment may be said to have been the principal ingredient in the fabrication of tales of this nature, and a little excuse is, therefore, to be made for men who, in deceiving the world may, it is just barely possible, have likewise contributed to deceive themselves. No such consideration, however, is due to the authors of another species of fabrication, which, originating in the same desire of self-aggrandizement, has been put forth as having occurred, without even the slightest foundation in truth, and in defiance even of their own knowledge and conviction of its impossibility.

The following example will best serve to illustrate the species of imposture here alluded to. I have selected it out of a great many of a similar description, because the adventure to which it refers is a very favourite one, both with the public and the profession, having been asserted by or attributed to almost all the celebrated aeronauts in succession. It was first related to me many years ago by the same gentleman, who was the subject of the marvellous act of regeneration before alluded to, (whose name for reasons there specified I have already declined mentioning,) as having actually occurred to himself; since which time I have heard it frequently repeated both in allusion to him and a variety of other persons; and that too, even in despite of their own most strenuous disavowal.

The adventure alluded to is supposed to have taken place on the occasion of a rent appearing in the silk of the balloon, while at a great elevation in the air, and which the individual is stated to have repaired by climbing along the network, and tying the opening with his pocket handkerchief.

To an ordinary or unreflecting auditor there is nothing about this incident calculated to produce any suspicion of its truth, beyond that which the difficulty of the proceeding, and the extraordinary intrepidity required to accomplish it, might be likely to awaken. To one, however, acquainted with the common principles of the art, or at all inclined to enter into a minute investigation of the affair, a variety of circumstances immediately suggest themselves, all, or indeed any one of which is sufficient to prove the whole a fabrication. It will, I think, afford some entertainment to the reader, as well as serve to illustrate a few of the leading features of the art, if we examine a little more minutely into the nature of the circumstances here alluded to.

In the first place then, one thing is pretty clear; namely, that in order to have given occasion for the attempt at all, the existence of the fissure must have been positively known to the aeronaut. Now such a discovery could not have been made by him *while in the air* unless the rent were in the *lower portion* of the balloon, that being in fact the only part visible to an individual seated in the car. It is true that certain phenomena, as for instance a rapid and unaccountable depression in the course of the machine, *might* have led him to conjecture that some such accident had occurred, in some part of the balloon beyond the reach of his vision. Upon such uncertain grounds, however, I hardly think any man in his senses would dream of resorting to so perilous and precarious an expedient, as that of attempting to mount the slippery and fragile sides of such a globe in such a situation in order to submit to personal inspection an extent of surface amounting to many thousand square feet, so small a portion of which, from the spherical nature of the object, would ever be visible to him at once. Nor indeed would he have time for such an operation, were he even disposed to attempt it. With such a rent in the most material part of the balloon, the escape of gas would have been such, that long before he could have concluded his operations, the balloon would have reached the ground; the more especially as it must have been in the act of descending, and that too rapidly, to have produced in his mind the suspicion of the accident and have impressed him with the necessity for taking the steps above-mentioned to arrest its progress. Were the rent alluded to therefore otherwise than in the inferior hemisphere of the balloon, the detection could not have been made by the aeronaut during the term of his flight.

Now, it is hardly credible that the knowledge of the existence of a rent which it could be thought necessary to repair at such a risk, could have been obtained *previous to starting*; as I hardly think any one would be likely to have deferred the operation of closing it, when it could be done with perfect ease and safety, until a period when so many difficulties and so much danger must have attended its execution. The probability, therefore, is strongly in favor of the supposition that the rent would have been in the *lower part* of the balloon. But it must be observed that an aperture in such a part of the machine is by no means a matter attended with much danger, or involving any very serious consequences either to the safety of the aeronaut or even the continuance of his flight. The gas, by its ascensive nature, always remains in the upper part of the balloon, unless when so much rarified by its elevation in the atmosphere as to occasion it to occupy and distend the entire of the silken sphere; in which case, so far from being detrimental, such an opening would be rather likely to prove beneficial, especially among the aeronauts of the "olden time," when the neck of the balloon was wont to be kept closed during the progress of the aerial voyage. Indeed, I have known persons (Mr. Green among the number), who have commenced their ascent with holes in that part of the balloon, cer-

tainly much larger than could have been closed by a
handkerchief, without either anticipating or experiencing
any considerable impediment to their proceedings. Such
an opening does, in fact, at all times exist in the neck
of the balloon under the best system of aerostation,
and no issue ever proceeds from it, except when the
balloon has become fully distended by the expansion of
the included gas; to answer which emergency, and supply
the place of the valve in case of accident or default in
the action of that part of the apparatus, is the special
object of its being so disposed. In the lower hemisphere
of the balloon, therefore, the occurrence of such an
accident would not be of serious moment; certainly not
such as to authorise the adoption of such perilous means
for its reparation: in any other part of the machine
its presence would, as we have already proved, have been
as much beyond his knowledge as the remedy would have
been beyond his skill. For the sake of further eluci-
dation of the question, we will, however, suppose that
the aeronaut was aware of the existence of such an
accident, either in the upper or lower surfaces of the
balloon, and in the former case mad enough, or in the
latter foolish enough, to attempt to repair it; still were he
to undertake the steps he would fain have us believe, a
variety of obstacles, by their nature *insurmountable*,
would effectually prevent him from accomplishing or even
approaching the object of his exertions. In the first

place, it will be observed that the net-work, by means of which he is supposed to have made his way, although, in its aggregate, sufficiently strong to support the utmost weight required of it, is yet composed but of slight filaments, far from sufficient to sustain the weight of a man in their separate and individual capacities. In the next place, this net-work, such as it is, does not reach down to the hoop, but when within about ten or twelve feet of the car (according to its size), merges into simple cords, which, tending towards the hoop as towards a common centre, form the connection between these portions of the machine. Along this interval, therefore, the aeronaut would have had to proceed ere he attained the netting, without further support, either for hands or feet, than what was afforded by the slender cords that lay stretched, in the direction of his progress, and nearly parallel, between them. Now, this wonderful progression, if ever it were to be done at all to any distance, must have been done on the *external* surface of the netting, and not on that side next the balloon. This is ascertained from the consideration of the great pressure (no less than the whole weight of the appended apparatus) exerted by the parietes of the balloon against the net-work which encloses it, to such an extent that it is very questionable whether any person could long exist, much less exercise his locomotive functions, in such a situation. On the *outward* side of the netting, therefore, it is clearly necessary that he should have proceeded. Now the balloon, as is well known, is a sphere of no small dimensions, and attached so near the car that the cords which connect them by means of the net-work and hoop start out from the latter at an angle, (when the balloon is inflated) not many degrees inferior to a right angle, and continue in that direction to a distance varying as the semi-diameter of the sphere; say, at the least, from fifteen to twenty feet. All this space then would the adventurer have had to traverse, *with his back downwards*, nearly horizontally extended, without accommodation for his hands or feet, beyond a few simple cords scarcely sufficient to support his weight, and in that awkward and constrained position proceed to perform an operation which, after all, would be both impracticable and insufficient to the end proposed, inasmuch as no rent of any magnitude, against which the whole pressure of an inflated balloon was exerted, could be effectually closed by the hand, or if closed, effectually confined by a pocket handkerchief.

In all this enquiry, it will be perceived we have gone upon the supposition that a certain line of proceedings, which it would have been necessary to adopt, would be attended by certain difficulties, which it would have been necessary to overcome. There is one consideration, however, to which we have not yet adverted, which would render all these proceedings nugatory, and by which all those difficulties would become at once converted into impossibilities. We all know that the balloon maintains its particular form and position in the atmosphere owing to the peculiar attachment of a certain weight. Wherever that weight be appended will always be the *bottom* of the apparatus, and *above* that weight and at such a distance as the connecting cordage will allow, the balloon must always remain. When the balloon is in that collapsed condition, in which it must have been with a rent like that in question, this distance will vary *at least* from ten to fifteen feet; an interval which must at all times and under all circumstances have ever continued to separate the aeronaut from the adjacent surface of the balloon. To advance along the net-work in the hope of reaching a more favorable station would be vain. Such an attempt would merely change the line of gravitation; the balloon, in the same form and at the same distance, would still hover above his head: like the squirrel in the cage, he would continue to advance without gaining a single step, or getting a whit nearer to the object of his wishes. All that he would effect would be merely to turn the balloon in the air, the spot where he clung to ever remaining the lowest, until having nearly reversed the machine, the former would most likely make its escape from the netting through the larger interstices of that portion of it which, when the balloon is in its proper position, is to be found in the neighbourhood of the car below.

Such then would be the result of any attempt to mount the sides of a balloon while poised in the atmosphere, and which proves the impracticability of the undertaking as clearly as we have already shewn that if it were feasible it would still have been inadequate to the purpose for which it was designed. Upon the whole, therefore, the case may be considered as an entire fabrication, and affords a very singular instance of the number of errors which may be embraced in one false statement. No less than *eight* circumstances, the majority of which are obstacles insurmountable by human means, appear in array to prove the impossibility of this seemingly simple assertion: viz. the ignorance of the occurrence of the fissure in a part of the balloon where its presence would require the operation alluded to; — the incapability of performing it in time to prevent the descent of the balloon;—the insufficiency of the net-work to support the partial disposition of his weight;—the distance he would have to traverse upon single cords before he would reach the netting;—the horizontal position in which he must have proceeded with his back downward, or the pressure (say no less than five hundred weight,) which he would have had to encounter in case he endeavoured to make his way between the netting and the silk;—the difficulty of performing the operation in the awkward position he must have occupied;—its inefficacy to produce the desired result even if practicable; and to crown all, the hiatus which, owing to the form and position of the relative parts of the machine impressed on them by the immutable laws of gravitation, must ever have continued to oppose an unconquerable impediment to the execution of his design.

It would be needless to multiply examples upon this score; enough has already been adduced to prove the utter disregard to truth which has generally signalised the contributions of the professors of aerostation, and consequently the little reliance which can be placed upon their assertions, more especially when the *mysteries* of the art, if we may be allowed the expression, are made the subject of their communications. To remedy this defect, in as far as in us lies, and display the phenomena of aerostation in their proper colours, is the object of the present appendix. For this purpose we shall endeavour to pourtray in the order in which they occur the most prominent of the impressions observable in the practice of the art, accompanying each with as correct an explanation of its origin and cause as it is in our power to afford.

The first thing then which strikes the incipient aeronaut in the outset of his career is the sense of extraordinary quiescence which immediately ensues upon the dismissal of the machine from the ground. No matter how agitated the balloon before its departure, no matter how violent the circumstances under which the ascent is effected, the moment the last hold upon the solid earth is

cast off, all is perfect repose and stillness the most profound. The creaking of the car, the rustling of the silk,
the heavy lurching of the distended sphere swayed to and
fro by the breeze, and shifting its load with sudden and
energetic motion, despite the efforts of the individuals
who are struggling to retain it, all have ceased in an instant, and are succeeded by a degree of tranquillity so
intense, as for a moment to absorb all other considerations,
and almost confuse the mind of the voyager from the suddenness of the change, and its apparent incompatibility
with the nature of the enterprise in which he is embarked.

Unprepared for such a result, or occupied perhaps in
other reflexions, the unpractised tyro is seldom in fact
conscious of the exact moment of his departure, and
instances are not infrequent in which the aeronaut has
been so far deceived by the unexpected serenity of the
situation as to have been transported to a very considerable elevation, without being aware that the act of separation had been effected, until it became forced upon his
notice by the fast fading voices of the assembled populace
cheering his ascent.*

* A remarkable instance of this occurred to Mr. Charles
Green in his first ascent, which took place on the occasion of
the coronation of His Majesty George IV. Oppressed with the
heat of the day and the fatigue he had previously encountered,
as he sat in the car waiting the sound of the gun that was to
indicate the moment of his departure, he requested his friends
who were holding the balloon to allow it to raise itself a little,

Recalled to the knowledge of his situation, a sudden and most natural impulse at first leads the aeronaut to look forward; nothing however appearing in the direction in which habit has almost unconsciously impelled him to direct his gaze, his eye insensibly assumes a downward course, and he becomes at once assailed with a mass of observations and reflections, among which, astonishment at the unusual tranquillity that accompanies alterations so rapid and so remarkable, is one of the most prominent. Without an effort on the part of the individual, or apparently on that of the machine in which he is seated, the whole face of nature seems to be undergoing some violent and inexplicable transformation. Insensible of motion from any direct impression on himself, and beholding the fast retreating forms, the rapidly diminishing size of all those objects which so lately were by his side, an idea, almost amounting to conviction involuntarily seizes upon his mind, that the earth with all its inhabitants had, by some unaccountable effort of nature, been suddenly precipitated from

that he might enjoy the fresh air above the heads of the crowd, that hemmed him in on all sides. In endeavouring to comply with his request the assistants accidentally let the cords slip from their hands. Disengaged from its hold the balloon immediately and rapidly commenced its ascent, without the slightest knowledge on the part of Mr. Green, who had just sat down to repose himself, and had actually reached a considerable elevation ere he was made sensible of the fact by hearing the united shouting of the multitude, accompanied by the expected discharge of the cannon, which almost miraculously took place in the same instant.

its hold, and was in the act of slipping away from beneath his feet into the murky recesses of some unfathomable abyss below. Every thing in fact but himself, seems to have been suddenly endowed with motion, and in the confusion of the moment, the novelty of his situation and the rapidity of his ascent, he almost feels as if, the usual community of sentiment between his mind and body having been dissolved, the former alone retained the consciousness of motion, whereof the latter had by some extraordinary interference been suddenly and unaccountably deprived.

Although the absence of all the ordinary effects of motion upon the human frame continues to mark the progress of the æronaut at all elevations, and throughout the whole of his career, the peculiarity of his situation in that respect is never so forcibly urged upon his notice as in the commencement of the ascent (when the contrast between his late and present condition is freshest in his mind), and, though in a slighter degree, during those depressions which occasionally take place in the course of the voyage, when the balloon happens to be brought into closer contact with the surface of the earth beneath.

I have already adverted (*page* 37,) in the preceding narrative, to the peculiarly delightful sensations that attend upon such situations, and among them have remarked as by no means the least striking, those which arise from the consciousness of rapid motion, unattended by those effects by which, in all other circumstances, it is ever known to he distinguished. No part, in fact, of the whole career of the aeronaut bears so strong a resemblance to flight, or more truly appears to realize the sensations we sometimes experience in our dreams, when elevated in fancy to the enjoyment of that delicious occupation. Here it is that the reality of the case is most strongly forced upon his notice, and the mind awakened by all the various symptoms of the rapid progression of the atmospheric current in which he floats—the sounds of its resistance, issuing as it were out of the very bowels of the earth—the agitation of the trees—the varying tints of the upper surface of the woods and meadows as they bend simultaneously beneath the blast—the rapid retrocession of all the known fixed and stable objects upon the plain beneath—together with the ever-changing features of the scene; all indications undeniable of the reality of his progress, which every foot he recedes from the vicinity of the earth only tends to weaken and impair. Truly conscious of his motion, here it is that he is most strongly impressed with the absence of its ordinary effects, and *feels* the novelty and delight of a situation which in no other manner can he ever be made to experience. As he rises, this feeling disappears, and he ceases to derive any extraordinary impression from the peculiarity of his situation, because, not being made sensible of the real state of the case from observation and reflection, he perceives no reason to suspect that there is motion, and consequently suffers no peculiar sensation or surprise from the absence of its ordinary effects.

That the body should thus, in a manner, be insensible to the effects of motion in a balloon, will not, perhaps, be deemed so surprising when we come to consider what are the means by which alone these effects are in ordinary cases made apparent to the human frame. As this is a new field of enquiry, for aught that I am aware of, the reader will excuse our taking a more minute review of it than, under other circumstances, we should perhaps feel ourselves authorised in hazarding.

In the pursuance of this enquiry, then, it is necessary to be observed, that the human body is composed of a variety of different materials, of different specific gravities, and endowed with different degrees of sensibility to pressure, or other disturbing causes, to which they may happen to be subjected. When these are set in motion all together, by one and the same impelling force, a very considerable disarrangement of their relative positions must ensue, or else a proportionably great resistance to that disarrangement, where the parts are so circumstanced as not to be able to change their position in obedience to the general impulse.

To make this clearer by an example, if a tray containing a variety of different sized globules of lead or other heavy material, varying in dimensions from a grain of sand to a four-pounder, be placed at one end of a long table or board fixed horizontally, and with a sudden motion be made to slide forward towards the other, a marked difference will immediately take place in their relative positions from that in which they were placed at first. The larger and heavier balls, unable to acquire the same rate of motion, in the same space of time, and through the medium of the same impulse, will immediately fall a little behind the others, and all, more or less in proportion to their particular *vis inertiæ*, suffer a retrocession or loss of place, owing to the suddenness wherewith the first principles of motion had been attempted to be communicated to them. Were these objects so disposed as not to be able to display the influence of the sudden acquirement of motion, by a change of place, (as for instance, if they were all connected together by elastic ligatures, or imbedded in glutinous strata), the motion thus impeded would necessarily resolve itself into a reaction among the parts, producing unequal degrees of pressure, or communicating unequal shocks (where any liberty for motion was allowed) to the adjacent portions of the medium in which they were located. Now, this is exactly the situation in which the parts of the human body exist, and which we have sought to represent in the previous example, by the more familiar illustration of the leaden globules.* Prevented by their structure and combination

* The reason for our selecting that material as an agent in the experiment, is merely on account of its weight, to avoid as

from following the course they would assume, if allowed to act in obedience to the laws of motion, all the motive energy with which they have been endowed is necessarily resolved into reaction, and being various in amount and variously exerted, produces a disagreeable pressure or tendency to disturbance of the condition in which the parts naturally exist when in a state of repose.

To this disturbance then, 1 am inclined to attribute the production of the sense of motion in the human frame, which may thus be considered as merely a new mode of operation in the sense of feeling, or rather perhaps of that sixth sense discovered by our celebrated physiologist Sir Charles Bell, by means of which the mind takes cognizance of the relative positions of the different parts of the body without the instrumentality of the organs of sight or feeling.

Now we learn by the laws of dynamics that all bodies, without regard to their specific gravities, move with equal velocities under the same active impulse in an unresisting

much as possible the influence of the resistance of the atmosphere in checking the tendencies of the different objects to follow the course pointed out for them by the laws of projectiles. In the application of the example to the human body, no such consideration is required, as all the parts united in one common mass are by their nature protected from all such interference, except upon their external surface; and even from that they are by the peculiar characteristics of the art, exempt in the process of aerostation.

medium ;* the only difference observable in their conduct being in the length of time required ere their powers of passive resistance be overcome, and they be brought to display the whole result of the motive force applied; as may be seen upon reference to the experiment which we have already adduced in illustration of the subject; wherein, after the first derangement of the relative positions of the objects on the tray; occasioned by the first induction of motion, no further derangement will be observable so long as the rate at which they are propelled remains the same. As it is upon this derangement alone that depends the sense of motion, one point in the train of consequences then becomes established, namely that no sensation will be awakened in any individual so long as the motion to which he is subjected is uniform.

Again, were those changes of motion, (to which we have above alluded as being the only causes of the derangements that awaken the sense of motion) to take place in such a manner as not to be productive of those derange-

* Although a vacuum and an unresisting medium are not exactly the same thing, yet as regards their influence in the laws of motion they may be considered as similar. The different internal parts of the human frame for instance are not seated " in vacuis," yet the influence which the medium wherein they are situated exerts upon them, disappears when *fairly* in motion, all the parts observing the same rate and therefore affording no grounds for interference.

ments, then would the epochs of those changes, like the others fail in being noticed, and the whole career of the individual, however varied, pass without the slightest consciousness of motion on his part. To this effect all that is requisite is the observance of a certain rate in the induction of those changes, whereby the *vires inertiæ* of the different parts are overcome, and all are made to commence their career of equal motion at the same time. By a slow and gradual process alone this may be accomplished ; for, however there is a limit to the quickness with which bodies will take upon themselves a given state of motion, there is no such limit in the opposite direction ; if you proceed to invest two unequal bodies with equal motions *too rapidly*, you will disturb their relative positions by investing the lighter with the full amount of motion, before you have entirely overcome the passive resistance of the heavier ; but if you proceed ever so *slowly* to the same end, you will never produce a derangement of their relative positions by investing either with the full amount of motion before the other. Accordingly, to resort once more to our favourite illustration, if the tray of objects above mentioned were to be advanced gradually and with proper regard to their several exigencies, the utmost conceivable rate of motion might be obtained, preserved, altered, abolished, and renewed *ad infinitum* without the slightest derangement in the relative position of the different component parts. It is almost unnecessary to add that were an individual placed in the same circumstances, the different parts of his body would observe the same laws and exhibit the same result; the consequence of which is, that under such circumstances, the sense of motion would not be awakened at all, and the irregular as the uniform progression pass equally unheeded and unknown.*

* The adoption of the preceding theory of the sense of motion will, I believe, afford a clue to the solution of certain physiological phenomena, which have long puzzled the world, and which, although not exactly pertaining to the present subject, yet, as being corroborative of the theory by which it is sought to be illustrated, we may perhaps be excused for noticing; I allude to the sickness experienced at sea, in the exercise of the swing, in the revolutions of the waltz and other movements of a similar description, and productive of similar results. From what has been said above, we perceive that the sense of motion and its immediate cause, the derangement of parts are not *always* attendant upon actual motion, but merely observable in consequence and during the continuance of certain interruptions. But the derangements alluded to, and consequently the sense of motion to which they give rise, are not capable of being excited to a very high pitch of energy by every species of interruption which may occur to call them into action. From the very nature of the construction of the human frame, these derangements of the parts can never without actual organic lesion take place to any very considerable extent; and consequently the sense of motion, as we really find to be the case, cannot be capable of great intensity. Like many other corporeal (and indeed all mental) impressions however, the deficiency in intensity of action, may be amply supplied by the protracted continuance of its effects: as an illustration of which in analogous cases, I need only cite the action of most medicines, for

Now this is exactly the situation in which the aeronaut
is placed. From the moment the balloon quits the ground

instance, that of the emetic principle, upon the stomach, which,
unaltered in its intensity, does not begin to act until the parts
have for some time been subjected to its influence. In the
same manner the derangements which give rise to the sense of
motion may be, and frequently are, by the increased duration
of their action, brought to exhibit very powerful and impressive
consequences. To produce that increased duration of action, it
is necessary that the sense of motion be supported by a course
of interruptions, occurring at such intervals as will not allow
the parts to recover from the effects of one deranging influence
before they have been subjected to another. That this is the
case as regards alternating motions, those for instance by which
sea-sickness is produced, does not require to be illustrated ;
the interruptions, by means of which the sense of motion is
maintained, are here sufficiently palpable. With respect to
rotatory motion, however, the action of the deranging causes
may not perhaps be quite so evident. Nevertheless, though
more obscure, they are not less decided, and if any thing still
more energetic in their effects. As all bodies in motion when
uninfluenced by disturbing causes, tend to proceed in right
lines, the motion of bodies conveyed in the direction of a curve,
may be considered as really compounded of incessant interruption to the rectilinear course, which the laws of nature incline
them to pursue. So far, therefore, from being exempt from
disturbance by the apparent equability of their motions, the
parts of a body revolving round a centre are even still more incessantly liable to the deranging agency than where they are
absolutely made to alternate, even with violence, between two
extremes in opposite directions.

To the *protracted duration of the sense of motion then*, I am
inclined to attribute all those cases where distressing symptoms

until its return to earth again, nothing ever befalls (except from accidental collision,) which can or does produce a change of motion sufficiently rapid to awaken the perception of his progress. Changes it is true do occur both in the rate and direction of his course. Alterations in his elevation are continually taking place with more or less rapidity, which occasionally require to be checked with

follow the infliction of certain movements. To this supposition all the phenomena are reconcileable. Here we see the reason why a heavy lurching motion, the heaving of a ship at sea, for instance, but still more, the rotatory motion (in which the disturbing influence is not only *protracted*, but *incessant*) is always attended by greater distress than a short, quick, alternating motion, however long continued, where the impetus of the parts becomes arrested before they have experienced the full amount of disturbance, and where, constantly oscillating on either side of their natural condition, they are never either *long* or *far* from the means of recovery. We also see the reason why in a rotatory movement the larger the circle in which the parts are conveyed, the less the distress; the tangent in which they tend to fly off more nearly coinciding with the segment of the curve in which they are detained. Thus, revolving rapidly on one foot, after the manner of the pirouette, is quicker in inducing nausea than performing the gyration in a larger space, to those who are unused to either. The manner also in which habit enables the individual to withstand the effects of the motions is also strongly in accordance with the principles of the above explanation, and might be illustrated by many analogies with other physical impressions, but that I fear I have already too long trespassed upon the patience of the reader in a matter which is foreign to the present subject, and in which, therefore, it is very possible he may not feel an equal interest.

considerable promptitude by a liberal discharge of gas and ballast; a few seconds are frequently sufficient to make a difference of some thousand feet in the level of his course; yet the changes, striking as they may be, are never accomplished with that degree of impetuosity which is necessary to awaken a sensation of their effects. Currents also of different velocities and different bearings are also constantly encountered. But the mutual action of currents of air is never sudden: their bounds are not fixed by certain lines, like those of the more solid substances, nor are the changes which may take place in them, even though conducing to direct opposition, ever so decidedly marked and promptly executed as to lead to a sensible perception of their results.

Debarred, therefore, in every way from obtaining a direct personal feeling of his progress, it is only by a comparison with the phenomena presented by known fixed objects that the aeronaut can even ascertain the fact, whether he is really in a state of quiescence or of motion. This is an intelligence which his sensations alone are incapable of supplying; it is to his judgment, with the assistance of his sight, that he is forced to look for the solution of the question. Where the exercise of that organ is denied him, as at night, during the prevalence of fogs, where clouds intercept his view, or the uniformity of the subjacent surface leaves him no sufficiently distinct object to refer to, as o'er a broad expanse of ocean, the rate of his progress, nay, its very existence, is to him a secret not to be unravelled, except by the aid of such a mechanical connexion with the earth as in his ingenuity he is able to devise. Such a connexion is that which is formed by means of the guide-rope; and the indications it affords of the rate and direction of the balloon, I consider not the least valuable property of that ingenious instrument.

The next striking circumstance in the order of succession, distinctive of the present subject, is the sudden cessation and continued absence of all atmospheric resistance, the presence of which one is apt to consider so essential a concomitant of locomotion, especially when conducted with any unusual degree of speed. Acting in conjunction with the preceding, the influence of this novel characteristic upon the mind and senses of the inexperienced aeronaut in the commencement of his career is truly magical; more especially if the state of the weather at the time be such as to afford room for the establishment of a sufficient contrast. Suddenly subsiding the instant the act of liberation has been effected, all the various symptoms of violent atmospheric commotion, by which his previous situation was so notably distinguished, simultaneously disappear; the heaving of the balloon, the sighing of the wind through the cordage, the flapping of the silk above his head, the wonted action of the passing breeze upon his own person, every thing, in short, which can bear testimony to the exertions of the element and the force by which it is with difficulty resisted, at once become completely sopite : not a motion is felt, not a breath of wind is perceptible; the balloon, as if arrested by the influence of some powerful and invisible agent, suddenly assumes an upright posture, and stands, as it were, fixed, rigid, and immoveable, while the mind of the adventurer, unconscious of all but the change itself, becomes struck with the awful conviction that some extraordinary revulsion has just taken place, whereby the raging elements have been suddenly tempered into tranquillity, and an universal and unnatural calm induced upon the previously disturbed condition of the mighty powers of nature.

From this time forward, until the conclusion of his flight, the same impressions continue to accompany the progress of the aerial voyager, weakened only in their energy (like indeed almost all those peculiar to the practice of this art) as, increasing his altitude he diminishes his relations with the earth, and with them the grounds of comparison, whereby alone he obtains a consciousness of the real circumstances of the case, and is made to *feel* the absence of results, which are in fact only remarkable when missed, and only missed when particularly expected.

So long as the balloon is left free to pursue her own course upon the same level, unaffected by any of those excessive variations in her buoyancy, which impress upon her a rapid motion, apart from that of the current in which she floats (as when she ascends or descends at the

commencement or conclusion of her career, or by the
sudden loss of any serious amount of gas or ballast during
its continuance), this state of things remains uninterrupted, admits of no qualifications, and is liable to no
exceptions. Totally independent of the rate or direction
of the current, it remains equally absolute whether the
actual progress of the balloon be one, or one hundred
miles an hour—whether it be on one continued line or
subject to the most rapid and incessant variation. The
greatest storm that ever racked the face of nature, is in
respect of its influence upon this condition of the balloon,
as utterly powerless and inefficient as the most unruffled
calm, the most unequivocal repose. To such an extent
is this the case, so truly indeed is atmospheric resistance
a nullity to the aeronaut, that were we to suppose him
(by way of illustration) suddenly transported to the West-
ern Indies, the birth-place and habitation of the tornado
and the hurricane, traversing the skies at a time when
one of the wildest and fiercest was exercising its utmost
powers of devastation, looking down from his air-borne
car and beholding houses levelled, trees uprooted, rocks
translated from their stony beds and hurled into the sea,
earth and ocean in mutual aggression encroaching upon
each other's limits, and all the various signs of desolation
by which its merciless path is marked, he might nevertheless hold in his hand a lighted taper without extin-
guishing the flame, or even indicating by its inclination to

one side or the other the direction of the mighty agent
by which such awful ravages had been created. No
sooner, however, has the grapnel touched the ground,
and the slightest opposition been afforded to the progress
of the balloon, than all this seeming tranquillity is at an
end, and the aeronaut for the first time becomes sensible
in his own person of the real influence of that mighty
element, of whose presence and power he had hitherto
been able to judge through the medium of his sight alone.

The theory, by means of which the non-resistance of
the atmosphere in aerial navigation is accounted for, is
by no means so complex as that by which it was found
necessary to illustrate the previous characteristic phenomenon of the absence of the sense of motion. Floating
in and by means of the action of the air itself, no difference can in fact ever exist between the rate of the
machine and that of the medium of its conveyance, (after
the first efforts to overcome the *vis inertiæ* of the former
have been successfully exerted,) so long as both remain at
liberty to obey the course dictated by the laws which
govern the motion of bodies in a fluid medium. Strictly
observing the same reciprocal positions throughout the
whole of their progress, no retardation or acceleration of
the course of the one beyond that of the other exists,
whereby a resistance could be created. To all intents
and purposes, therefore, a balloon freely poised in the
atmosphere may be considered as absolutely inclosed or

imbedded in a box of air; so completely so, that (for example) were it possible to distinguish, by tinging it with some particular colour, that portion of the atmosphere immediately surrounding the balloon, and in that guise commit her to the discretion of the elements, she would, apart from all fluctuations in the level of her course, continue to bear the same tinted medium along with her, even until having completed in her course the circumference of the globe, she had, the winds permitting, returned to the same spot from which she had originally departed.

As a general rule, however, it is to be observed, that this characteristic discontinuance of atmospheric resistance only holds good as regards the horizontal or *passive* progress of the balloon. With respect to its vertical, or as it may be termed, its *active* motion, that in short which proceeds from the exercise of its own buoyancy, some deviation from that state of perfect atmospheric repose will no doubt be occasionally discernible, especially when the movements alluded to are accomplished with any considerable degree of rapidity. Upon the principle of this resistance, various attempts have been made to construct instruments to supersede the barometer, in affording indications of these movements, and of the rate at which they are effected; hitherto, however, it must be confessed, without any satisfactory result. The generality of the changes are, in fact, much too slowly conducted to afford grounds for the establishment of a resistance sufficient to overcome the obstacles offered by the *vis inertiæ*, friction, defective construction, and " the thousand natural ills which *art* is heir to," and from which no species of instrument, however delicate, which depends on mechanical action for its results, is entirely exempt.*

From what has been before stated, the futility of any attempt to apply this principle to the ascertainment of the horizontal motion of the balloon, either by means of instruments especially constructed, or by observations drawn

* The best of these attempts which I have seen is undoubtedly that of Mr. F. Gye, Jun., (son to the proprietor of the balloon in which the late expedition was accomplished) upon the principle of an extremely light wheel adapted with vanes, like the paddles of a steam-boat, and inclosed in a box partly open at top and bottom, to admit the action of the air in ascending or descending. To the above is attached a rotary index, serving to denote by the velocity and course of its gyrations the rate and direction of the machine " in transitu." Although the result of the trial to which it was submitted in our excursion was not perfectly satisfactory, it is but just to observe, that the fault was more attributable to the defects of the particular instrument than to the principle of the contrivance; its size being too limited to take in a sufficient portion of the atmosphere, while at the same time it was not sufficiently protected by the form of its receptacle from the influence of the counter-currents, occasioned by the motions of the larger body in its vicinity, whereby its action in the former case was impaired, in the latter, deranged. With a due consideration of these defects, the result would, I have no doubt, have been more favourable, though never to such an extent as to enable it to supply the place of the barometer.

from the difference between the rate of motion of the balloon itself, and that of light bodies (as tissue paper for instance), purposely thrown over, is placed beyond a doubt. No such difference, in fact, occasioned either by the detachment of the body or its different specific gravity at all exists: where any such is perceivable, or thought to be perceivable, it may at once be laid to the account of some peculiarity in its form, or otherwise in the direction first impressed upon it, and which in the course it induces it to assume, is as likely to have acted in opposition to, as in concert with the direction of the current at the time prevailing.

Bound, of course, by the same rule, all clouds occupying the region of the same current in which the course of the balloon happens to be conducted, must ever continue to observe the same distances from that object as they held at the commencement. It is true that internal changes of form and position may at all times be discerned between the different parts of the same vapoury stratum, by any one who will take the trouble to examine their progress attentively for a few minutes. Without, however, infringing upon the generality of the proposition here laid down, such alterations of form and position are amply accountable for on the score of temperature, electrical affinity, and a variety of other specific influences; either through their direct effects upon the forms and dimensions of the aqueous masses (and be it observed that a change in form is in fact a change of position too), or by reason of the variations they work in the actual densities of the different parts, whereby their existing momenta become altered, and a temporary interruption occasioned in the equability which (with such exceptions) characterises their motions, as that of all other bodies, in an unresisting medium.

The entrance therefore into clouds and the exit from the same can never take place without a change of altitude on the part of the aeronautical machine ; an observation which may give some satisfaction to those who rate highly the danger of coming in contact with clouds charged with electric matter, or entertain a fear of being overtaken by bad weather in the course of their excursions.

One other consequence of the absence of atmospheric resistance remains to be noticed; I mean its influence in mitigating the effects of a low temperature upon the human frame, and rendering regions not only habitable but even delightful, which, but for this modification could never be entered without pain nor endured without danger. In the previous narrative, (page 70,) I have already adverted to this circumstance, and noticed the beneficial consequences that resulted to us from it during a night and a voyage of otherwise insufferable rigour. In further illustration of the effects of that peculiarity to which I have attributed the exemption we experienced from the

ordinary consequences of a low temperature, I have merely
to add that the only periods when the actual temperature
pressed severely upon our feelings were, when in ascending or descending rapidly, as occurred to us occasionally
during the night, a motion and resistance was occasioned
in the air, such as I have just mentioned to be the only
exceptions to that general state of atmospheric stillness
which otherwise never ceases to distinguish the progress
of the balloon in her career. What dependance is to be
placed upon the statements of those aeronauts who have
indulged in such glowing accounts of their sufferings by
cold, when, in pursuit of their vocation, they aimed at
penetrating into the higher regions of the sky, we shall
scrutinise when we come to that part of the present appendix which professes to treat of the phenomena observable in such situations.

To return to the aeronaut whom we left some pages
back in the act of commencing his ascent, the reader
must not suppose that all the circumstances and impressions
which we have here detailed as consequent upon the
change which the liberation of the balloon is calculated to
make in his situation, or the same tedious process of reasoning by which we have found it necessary to explain
them, are either adopted or even recognised by the individual at that particular epoch of his voyage. It is not,
indeed, at the time, certainly not the *first* time of experiencing them, that the aeronaut ever becomes awake to

the just amount of his feelings, or fully conscious of the real combinations to which they are to be attributed. Indeed, to arrive at the latter of these conditions, requires a course of analytic examination to which few persons have sufficient presence of mind, or rather *insensibility to the charms of their situation* to be able at such a moment to submit; and even were they so inclined, the celerity wherewith the first operations of the ascent are conducted, and the variety of the events and sentiments by which they are succeeded, are such as to leave no time for the consideration of any one in particular, unless to the utter exclusion of all the rest. It is by the frequent experience of the enjoyment alone, or the constant recurrence to it in after times, through the medium of the recollection, that a thorough knowledge is obtained of all its various peculiarities, the effects of which are much more generally experienced in the mass than in detail, and, by most persons at least, much more readily acknowledged than understood.

From commenting, therefore, upon the state of his own feelings, the attention of the aeronaut is early and forcibly recalled to a consideration of the " world without him," where, indeed, a new and fertile source of gratification awaits him, in the prospect which his increasing elevation has almost unconsciously presented to his view. No sooner, in fact, has he cleared the highest obstacles in his immediate vicinity, ere he finds himself apparently in the midst of his career, and hurried into the presence of all those objects which constitute alike the study and delight of the aerial voyager. Indeed, the celerity with which the translation is accomplished, and the curious conclusions to which it conduces in the mind of the beholder, are not the least striking circumstances of the whole proceedings. Springing as it were at a bound out of the very bowels of the earth, scarcely a second elapses ere the balloon, approaching to all appearance the very acmé of her ascent, has placed the astonished beholder in full view of the spectacle prepared for him; not as it were with one sudden stride, or at one unvarying velocity, but seemingly like a rocket shot from its frame, that with decreasing energy continues to mount, until, at length, its utmost force being spent, it appears to pause for an instant ere it turns to bend its downward course to earth again.

Such, in fact, is the impression which the circumstances of the case are most strongly calculated to produce upon his mind, and from which nothing but a perfect knowledge and firm conviction of the reality could effectually preserve him. Without the sense of motion to guide his judgment, the only opinion he can form of his ascent is necessarily though unconsciously drawn from a hasty consideration of the changes which it occasions in the aspect of the scene around him. Now as by the nature of things, all these changes proceed with *rapidly diminishing* intensity, as

the distance from the eye of the spectator becomes increased,* so under the same condition of removal must the

* The linear dimensions of objects being determined by the
angle under which they are seen, necessarily vary in the inverse
ratio of their distances from the point of sight. By the same
rule it follows, that the superficial dimensions, upon which their
apparent sizes depend, must vary inversely as *the squares* of
the distances from which they are beheld. Thus, a body seen
from any given point would appear four times as great as if
seen from twice the distance, nine times greater than it would
appear from a distance of three times the amount, and sixteen
times as great as if the eye beheld it from a position at four
times the original distance.

If in the place of the *proportionals* here employed to designate
the progression of the apparent decrease at stated intervals we
were to substitute *absolute* numbers, and estimate the dimensions
of the object as seen from a given altitude, say one hundred
feet, at the value of one hundred and forty-four, were the eye
of the spectator removed to twice the distance, or to an elevation
of two hundred feet, the number which would represent its
apparent magnitude would be but thirty-six, thus showing a
difference of one hundred and eight degrees between the appearance presented by the same object at the two stations in
favour of the former. Were, however, the eye to be still further
removed, to an elevation of three hundred feet, (being an in-
crement *equal* to the previous one), the measure of its appearance would be sixteen, thus denoting a loss of only twenty
degrees upon the second progression; while nine being the
expression of its visual magnitude at the height of four hundred
feet, would indicate a difference of only seven degrees lost
during the process of its removal through a third interval, equal
in amount to either of those which preceded it. In such a
series as this, it is unnecessary to observe, that an elevation is

sentiments of his progress in the mind of the aeronaut become continually impaired, until at last the alterations from distance having soon ceased to be appreciable, the sentiment of his removal, to the maintenance of which they alone had contributed, become alike rapidly extinct.

The case is one to which nothing analogous exists in nature or can be created by the ordinary exertions of art; consequently the effects and impressions to which it gives rise are such as can never be experienced but in a like situation and under exactly similar circumstances. In no other manner is or can the individual be abstracted from

very soon attained where the differences occasioned by equal increments of altitude, become so minute as to be inappreciable by the ordinary exertions of the senses. Now, as the impression of his ascent in the mind of the aeronaut, (deprived, as we have shown him to be, by the peculiar circumstances of the case, of all absolute sense of his translation), is entirely founded upon and regulated by these, the ocular effects of his removal, it follows that all personal knowledge of his ascent must rapidly and progressively become fainter, till at first hundreds and finally thousands of feet pass unnoticed, at least as far as the eye is capable of judging by a consideration of the altered aspect of the objects it surveys. Hence the difficulty of ascertaining the vertical direction of the balloon's course by the mere intervention of the sight alone, and the inestimable utility of the barometer in affording indications of the many changes which are constantly taking place in the level of her progress, and which in default of such indications would otherwise be unobserved until perhaps too late to remedy them without inconvenience.

the community with other objects of the same known appearances whereby to regulate his judgment and confirm
his conclusions. The situation which approaches nearest
to it in its conditions and effects is that of the mariner
when launching into the broad bosom of the ocean he looks
back upon the shores he is quitting, and beholds them
gradually disappearing in the obscurity of his increasing
distance. Even here, however, the objects are necessarily so limited, and the first steps of the progression (in
which the whole of the effect is concentrated) compara-
ratively so slow, that the alterations produced are too few,
and what there are of them too slight to afford grounds
for the institution of a comparison between the two cases.

Under the impressions we have here feebly endeavoured
to explain, and which time can neither obliterate nor
practice entirely overcome, the aeronaut quits the earth
to assume a station in the zenith of his own horizon. In
a few seconds all those capital changes by which, as I
have just stated, the first proceedings of the ascent are
invariably accompanied, have subsided ; and the prospect
has become sufficiently composed to admit the minuter
contemplation of its contents.

There projected upon a plane at right angles to his line
of vision, the whole adjacent surface of the earth lies
stretched beneath him, affording an heterogeneous display of matters at once the most interesting and incon-
gruous. Distances which he was used to regard as im-

portant, contracted to a span; objects once imposing to him from their dimensions dwindled into insignificance; localities which he never beheld or expected to behold at one and the same view, standing side by side in friendly juxtaposition; all the most striking productions of art, the most interesting varieties of nature, town and country, sea and land, mountains and plains, mixed up together in the one scene, appear before him as if suddenly called into existence by the magic virtues of some great enchanter's wand.

It is not, however, to the objects alone, magnificent and interesting as they may be justly deemed, so much as to the modifications they undergo from the unusual manner in which they are viewed, that is mainly attributable that peculiar effect by which the terrestrial landscape is so notably distinguished in the estimation of the aerial admirer. Seen, in the first place, from above, every thing that meets his eye, meets it under a novel aspect, and one which no other situation can in like manner and to the same extent enable him to enjoy. The summits of mountains, the tops of buildings, the upper surfaces of woods, those parts, in short, of all objects which by their natural or artificial positions, have hitherto been excluded from his view, are now almost the only ones that come within the scope of his observations. Indeed I can hardly conceive a prospect more interesting both from its novelty and the exquisite impressions to which it is calculated to give rise than that which a richly wooded and irregular tract of country presents when examined from the car of a balloon, either suspended motionless or slowly advancing within a few yards from the level of its upper surface; such a scene and such a situation for instance as that enjoyed by us when we found ourselves unexpectedly becalmed above the woods, after our first ineffectual attempt to take the ground at the termination of the expedition which forms the subject of the preceding narrative.*

The large, rounded masses of soft, green foliage, following generally the character of the subjacent soil, here swelling into mounds, there subsiding into hollows, altogether presenting the aspect of a mighty sea of verdure; sometimes intersected with roads or paths; occasionally opening to expose small portions of the groundwork, patches of mould, or little recesses of a more sparing vegetation; flocks of birds roused from their engagements by the unwonted intrusion upon realms, hitherto entirely their own, flying from place to place in the vain hope of escape, first in a body, and finally as the balloon tops the agitated community, breaking asunder and dispersing in every direction over the surface of the earth; the alternate approach and retreat of the beholder in connexion with the ground below, occasioned by the superior extancy of the hills, or the unusual depression of the valleys, introducing

* See page 79 *et seq.*

to parts otherwise inaccessible by human means; these and a thousand other circumstances and effects of minor note and less availing influence, combine to form a scene of enchantment in which the place of the sublime is amply supplied by that of the beautiful and the picturesque. Nor does it perhaps conduce least towards the general effect of such scenes, especially when viewed from a superior elevation, that all the objects of which they are composed are presented to the eye in the simplest manner possible as to their relative positions. None of the usual interference of parts, by means of which alone their different stations upon the same horizontal surface become assignable, is here to be perceived; nor any of those apparent variations in their dimensions which mainly serve to indicate their progressive removal from the point of sight, when situated in or about the same line of visual observation. All the ordinary qualifications of such scenes become, in fact, annihilated, and the eye for the first time beholds a picture of nature on the vastest scale, both as to size and magnificence, in the construction of which none of the complicated laws of linear perspective are at all involved.

As the balloon continues to ascend, another scenic peculiarity begins to display itself in the vividness of contour, the remarkable sharpness of outline by which the different features in the terrestral prospect are qualified, and which, strengthening with the increasing distance, never forsakes them so long as the objects themselves continue to be distinguishable. The roads, rivers, canals, streets, buildings, inclosures, hedges, furrows, watercourses, and all the various characteristics of rural and artificial scenery, instead of appearing obscured and rendered more indistinct by their remotion from the point of sight, seem on the contrary to augment in clearness and decision, and absolutely gain in intensity what they lose in the magnitude of their proportions.

This singular property is attributable to two circumstances, the union of which is another peculiarity of the art we have taken upon us to illustrate, namely, an increase of distance between the objects and the spectator, attended by a corresponding decrease in the density of the medium through which they are beheld; whereby the minuter features of the lines by which they are bounded, (and on which the irregularity of their appearance depends,) are exclusively lost to view, the objects themselves remaining as clearly distinguishable as ever. The process by which this conclusion is attained is very simply explicable on the grounds of the difference between the optical effects of absolute *remotion* from the point of sight, and those of mere *obscuration* upon the visual condition of the material world. Although the end to which they both conduce may virtually be the same, namely, the exclusion of the object from the view, yet their modes of operation are extremely different, and during their continuance give rise to very different phenomena. The indistinctness which the increase of distance, *per se*, occasions in the aspect of an object, is the consequence of its apparent *diminution*; while that which proceeds from the obscuring tendency of the medium through which it is beheld, is the result of a *concealment*, more or less partial, in proportion to the density of the said medium or the quantity of it which intervenes. By the former, the objects or the parts of objects are abstracted from observation *in the order of their several sizes*, commencing with the smallest; by the latter all are simultaneously and equally affected without regard to their dimensions. Now sharpness is a condition of the outline depending entirely upon the apparent absence of all parts bearing a small relative proportion to the whole; that which therefore removes from the sight such parts exclusively, conduces towards the production of the condition in question; and such an agent is distance, taken abstractedly. An antagonist to this result under ordinary circumstances, however, exists in the general indistinctness which ensues upon the quantity of the atmospheric medium in its *greatest density*, which is made to intervene by the very act of removal; so that before the beholder has sufficiently increased his distance from the object to enable him to lose sight of its irregularities, either the object itself has entirely disappeared, or so forfeited its general character of distinctness that no definite outline can be at all perceived.

From the influence of this interference, however, the aeronaut is to a considerable degree exempt; looking in the direction of the least atmospheric amount, he not only beholds every thing through the smallest possible quantity of obstruction consistent with his distance, but keeps constantly adding to his advantages in respect of the former, the more he continues to amend his position in respect of the latter. If the reader has ever during the prevalence of general fine weather, observed the aspect of some distant line of mountain, just before the occurrence of an unexpected shower, and noticed the peculiar clearness it appears on a sudden to have assumed, he will have witnessed a state of things similar to, though much weaker in their effects, than that which we have here attempted to describe; wherein the temporary rarefaction of the atmosphere, (the ordinary precursor of rain), acts the part of the vertical elevation of the aeronaut in reducing the quantity of intervening medium, and in paving the way for a readier admission of the distance to perform its share of the effects before attributed to it.

As soon as the adventurer has sufficiently recovered from the influence of these, the first and most predominant impressions, to be able to direct his attention to the other peculiarities of his case, he becomes gradually struck with the extraordinary degree of ease wherewith he feels himself able to regard his situation, and the total absence of all those sensations of giddiness and mental anxiety which he has always felt and conceived inseparable from positions apparently analogous to that which he at present occupies. Instead of shuddering, as he might fairly be supposed inclined, at the prospect so unusually placed before him; instead of drawing back, as it were, into himself to escape the full acknowledgment of the precariousness of his situation, he is astonished to find himself intently poring over the new leaf in the book of nature, which triumphant art has just enabled him to peruse, and far from trembling at his contents, enjoying in perfect tranquillity of mind the wonders it is continually unfolding to his view.

Nor is this a privilege by any means restricted to solitary cases, or dependant in any way upon the physical or mental constitution of the parties by whom it is experienced. All sorts of persons of every age and sex, and with every imaginable distinction of character endowed— the bold and the faint-hearted—the strong and the weak— the healthy and the infirm—equally concur in acknowledging the exemption; nor have I ever either met with or heard of any one of the numbers who have hitherto made practical trial of the fact, that ever complained of having been afflicted with the slightest giddiness or sense of personal anxiety from their exposure to a situation which, in the commencement at least, must have been equally unusual to them all.

From the earliest ages of the art, and even still (though owing to its more extended practice, in a less degree), this peculiar exemption has ever formed one of the sources from which the practical aeronaut has drawn most largely for his credit and estimation in the eyes of the uninitiated and admiring public. And, indeed, where the real state of the case was confined to the bosoms of the few, and the world remained in ignorance of the fact that the situation in question was as perfectly exempt from all the causes, as it is from the effects of those sensations in apparently similar cases so alarmingly experienced, it is no wonder that men should visit with an extra degree of admiration those who were supposed exclusively to have had the courage to defy and the fortitude to resist the assaults of feelings which, in their real presence, prove superior to every exertion of human nature, and, unless when conquered by long habituation, subdue alike the powerful and the weak. A very pardonable desire to make the most of such a peculiarity, has accordingly induced many aeronauts to make their ascents under circumstances of exposure particularly calculated to enhance the apparent dangers of the case and afford room for the exercise of such apprehensions in their fullest force, were they at all capable of being felt in such situations. One of those, a French aeronaut, M. Mosment, (whose fate, owing indeed to this particular circumstance, we have recorded in a subsequent part of the present volume*) was in the frequent habit of ascending upon a simple platform, entirely devoid of any defensive apparatus whatever—a practice in which he has been followed by many others, though happily for themselves, without experiencing so unfortunate a conclusion. I remember in one instance a brother of Mr. Charles Green, (Mr. Henry Green, likewise an aeronaut of considerable notoriety), when from some deficiency in the process of inflation he was unable to procure gas enough to raise the weight usually required, after rejecting in succession every article which could be conveniently spared, without succeeding in obtaining the necessary buoyancy, finally cut away the car itself, and assuming a seat upon the bare hoop, in that guise not only acquitted himself of his undertaking, but proceeded to a very considerable altitude ere he concluded his career.

Amongst the most remarkable instances of disregard in such particulars are, however, those afforded by the ascents of persons on the backs of quadrupeds attached to the balloon in the place of cars, and without other support than that supplied by the simple cords by which the animals were fastened to the hoops. I have seen several prints descriptive of performances of this kind, accomplished by

* See the memorandum affixed to his name in the alphabetical list contained in the Appendix c.

the French aeronaut, M. Margât, mounted upon a stag ;
one of which came off at Lyons, at what time, however,
I am not exactly aware. On the twenty-ninth of
July, 1827, Mr. Charles Green announced an ascent
to take place from the Eagle Tavern, City Road, on
the back of a favourite pony, which he had especially
trained for the purpose. The animal was in the first
place provided with two straps, loosely carried under
the belly, and made fast at the four extremities to corresponding situations on the hoop. Beneath him was a
small platform or tray, adapted with four receptacles for
the insertion of his feet, on which the whole weight of
his body really relied ; the straps before-mentioned being
merely intended as a precaution to prevent his falling,
in case from fear, fatigue, or otherwise he should feel
disposed to decline the standing, and endeavour to take
refuge in a recumbent position. In this fashion, Mr.
Green having mounted upon him in the usual way,
the ascent was accomplished, and they rose to a very
considerable altitude without the pony exhibiting the
slightest symptom of alarm ; on the contrary, indeed, while
at their greatest elevation the animal continued to display
all his wonted playfulness, and ate freely of a quantity of
beans, with which his gallant rider from time to time supplied him, from his hand. After a pleasant excursion of
some hours, they both descended in perfect safety near
Beckenham, in Kent, where they were hospitably enter-

tained by the neighbouring gentry.* This experiment has, I believe, been since repeated by Mr. Green with like success. These examples, with many others which might be adduced if necessary, serve to show to what an extent the peculiar immunity in question really exists, and how small a share the mere circumstances of pro-

---

* The pony in question is just as deserving of a biographical notice as many of the personages who figure in the annals of aerostation. I cannot exactly say to what particular district he owes his birth, nor what may have been the particular condition of his parents. His education, however, was by no means neglected, nor were his younger years passed in the obscurity to which so many of his kind have been devoted. He was, in fact, at a very early age, bound apprentice to Mr. Ducrow, and for some time figured at Astley's amphitheatre in the very first characters, till a change of circumstances and the rage for foreign novelties, which has proved the bane of the dramatic world elsewhere, drove him from the regular stage, and forced him to seek subsistence as an itinerant performer in a party of gypsies: in which capacity he continued for several years till accident brought him acquainted with kinder friends, and united him in bonds of friendship and humble dependence with Mr. Charles Green. Here he commenced a new career, and for some years contributed to adorn the skies as he had before the earth. Finally after having experienced more *ups* and *downs* than any horse, perhaps, that ever existed, he quitted a life of public service, and was buried in the garden of his master at Highgate, where he now reposes. It has lately, I hear, been designed to have his body exhumated, and forming a skeleton of his bones, in default of worthier materials, make of himself the monument of his own glory. In the mean time, we owe this tribute to his memory.

tection against the fear of falling, contribute towards its establishment.

Why the elevation to so unwonted an excess by means of the balloon, should not be attended with, to say the least, an equal degree of giddiness to that experienced when standing upon an eminence on the immediate surface of the earth is a circumstance which has been much canvassed and variously accounted for. By the majority of those who have considered the matter, this singular privilege has been supposed to be owing to the want of a visible connexion between the earth and the balloon, whereby the eye is precluded from measuring mechanically, and the judgment from painfully criticising the altitude to which the individual has been raised. That the want of a connection is the agent by which the result in question has been wrought, I have no doubt; as this, in fact, is the only characteristic distinction between the two situations; but that the mode in which it is said to operate is not the true one is pretty evident from the fact, that there are many situations which observe the same condition of a want of visible connexion with the earth, where the sensations in question are nevertheless found to prevail with unmitigated severity; as, for instance, in standing upon the summit of the monument of London, from whence all view of the pillar itself is excluded by the peculiar projection of the parapet; while on the other hand, situations fraught with an equal degree

of apparent danger abound, in which the connexion in question is amply discernable, without in the least contributing to excite a sentiment of his danger in the mind of the
individual exposed to it; as, for example, when he stands
upon a narrow plank, or bridge, firmly extended between
two perpendicular eminences, like that generally known
as the "Pont du Diable," in Switzerland, and from
which all apprehension of falling over has been removed
by the presence of a sufficient protection in the form of
a balustrade, or breast-work.*

From these examples, then, we clearly ascertain that
the mere absence of a visible connection is no more
available to prevent, than its presence is to occasion, the
production of the sensations alluded to, in circumstances
otherwise calculated to encourage or suppress them.
But the truth is, that the mental process of comparison,
to the want of which the aeronaut is supposed to be
indebted for his especial freedom from personal alarm,

* Another proof of the influence of the condition of the fulcrum in modifying the sensations in question. The tranquillity
experienced in the above situation is merely owing to the satisfaction in that respect which the support of the bridge at both
ends is calculated to afford. Were the bridge a projection supported at one end only, there is no question that, however convinced of its security by an examination of the nature of the
material, and its construction, the full force of the sensations in
question would be experienced, in despite of the sense of protection which the balustrade is otherwise competent to produce.

can have really little or nothing to do with the condition of his case in that particular. It is not, by any means, in proportion to his elevation that the sensations in question display themselves; nor indeed beyond a certain point does it seem to operate at all: the same impressions being consequent upon a station on the top of an ordinary house of five stories, and one upon the summit of the cupola of St. Paul's Cathedral, so far at least as the question of altitude is concerned. All that is required is, that the distance be such as to satisfy the mind that vital injury would accrue from the fall were it to occur. Now *that* knowledge it obtains without the aid of any visible communication with the earth; consequently it could never owe its exemption from the sensation in question to the want of a condition, of which if it were present it would never have availed itself.

The process, therefore, by means of which the deficiency of connexion in the case before us conduces to the admitted result, is unquestionably different, and the difference I take to consist in the light in which it disposes the mind to regard the security of the sustaining power. In all situations in which grounds of apprehension exist, and the apprehensions themselves ensue, a sense of personal insecurity may be decidedly affirmed to be the main-spring of their existence, the point upon which they hinge, and by which, in their continuance and amount, they are entirely and involuntarily determined.

Now as there are but two casualties by which the personal safety of the individual so circumstanced, can be compromised, namely, the loss of his equilibrium, and the precipitation by his weight of the fulcrum on which he relies, it is clearly to the involuntary dread of one or other of these two events, or the combined agency of them both, that the sensations themselves are to be ascribed, and of the nature of which, in quality and amount, they may be said in a manner to partake. Both these causes of alarm, however, are perfectly distinct, and, like the sensations to which they give rise, capable of acting either separately or in concert, according as the particular circumstances of the case may incline. How completely the exemption from any grounds of alarm on the score of the latter of these, (the apprehended instability of the sustaining power), is inadequate to save the individual from experiencing the full force of the impressions in question, while his condition with regard to the former (the insecurity of his equilibrium), is such as to give sufficient cause for their presence, it is unnecessary to demonstrate, both because the position is sufficiently evident without it, and also because the argument to which it tends is not needed in the illustration of the present question. That the security of the individual, in respect of the retention of his equilibrium, is no bar to the prevalence of the sensations in their fullest force, whenever the situation in other respects is qualified to call them into action, is, however, more to our present purpose, and though perhaps not so generally admitted, not the less true; as may be proved by any one standing upon the brink of some parapeted eminence, the whispering gallery of St. Paul's, or any other situation alike precipitous and yet protected from the danger of falling over; or when extended at full length, he endeavours to peer over the edge of some steep declivity; all positions from which the possibility of losing the equilibrium is removed, and the apprehensions of insecurity completely transferred from the individual himself to the fulcrum upon which he rests. From the consideration of these facts, taken in conjunction with the numerous examples we have already detailed, wherein even the ordinary defences of the art have been with perfect impunity dispensed with, we ascertain one important point in the train of our investigation, viz. that it is not to the peculiar construction of his vehicle, and the protection it is calculated to afford against the dangers of falling out, that is in any way to be ascribed the remarkable freedom of the aeronaut from the rigour of those impressions to which his situation in other respects one would be disposed to imagine above all others especially liable. Indeed, the share which his advantages in that particular can have in determining the singular tranquillity of his mind could never be of any very great importance, inasmuch, as, after all, the danger arising from this quarter is but of a minor note, compared with that occasioned by the insecurity of the sustaining power. The one is to a certain extent dependent upon the individual himself, and *may* be overcome by strong exertion, long habit, and particular constitution; the other is a casualty entirely beyond his controul, against which no exertion of his own is available to protect, and to which no habitation, however extensive, can in the least reconcile or inure him.

Were there grounds for apprehension, therefore, in any way imputable to the condition of his sustaining power, it is clear that the circumstance of his situation in other respects would never have been available to their suppression; a satisfactory evidence, therefore, that none such at all exist. To what then are we to ascribe the singular exception to the usual rule, in favour of the power by which the aeronaut is upheld? or in what manner does the want of connexion, which is its only peculiarity, contribute to the establishment of that immunity which it pre-eminently confers above all other situations, to which any shadow of danger is at all attributable? Simply by the manner in which it removes from the mind all the ordinary causes of alarm, and disposes it to admit without hesitation the assumption of its complete security.

As long as the circumstances upon which the fate of an individual depends, are such as to awaken in his mind a doubt of their competency, a tranquil sufferance of his condition is entirely out of the question. The influence of uncertainty, at all times in cases of personal alarm, more painfully insupportable than the actual presence of the thing apprehended itself, is nowhere more strongly manifested than in situations of the nature of those at present under consideration. The bare suspicion, that the fulcrum upon which he relies is about to break away and fall from under him, when once raised in his mind, is an idea so replete with horror that nothing short of absolute conviction, acquired through the evidence of his own senses, is capable of producing confidence sufficient to enable him to bear his situation with any thing like equanimity or satisfaction. It is of no avail to the pacification of his fears that any one should remind him that the brow of the eminence upon which he stands in fear and trembling has borne the brunt of ages and the weight of hundreds, or that the lofty column from behind whose guarded battlement he can scarcely persuade himself to look forth is really secure, and that its perdendicularity, from which it appears to him to be in the very act of inclining, is a condition much too stable to be cancelled by the weight of a single individual; so long as his senses continue to indicate a *possibility* of the occurrence of what he dreads, the assurance, nay, the knowledge, of its *improbability* is quite insufficient to neutralize their evidence, and overpower their suggestions. Indeed, the process of reasoning is an undertaking far too elaborate for the occasion, even were the individual disposed to encourage it. In situations of such impending physical peril, the mind has neither time nor calmness sufficient to enter into a calculation of chances, or to balance the arguments in favour of destruction and those against it, with a view to being guided by the result. The consequences of the conclusion are much too important, and if unfavourable, far too terrible, to be weighed for an instant; and the mind at once rejects with horror any attempt to reconcile it to a situation which allows of the chance of an issue fraught with such irreparable mischief, and teeming with distress even in the very thought. From all these painful impressions, nothing but a conviction of his security can ever entirely relieve him; a conviction obtainable only through the exercise of his powers of sight. Any tendency towards *concealment* on the part of the power by which he is sustained, operates to an enhancement of his anxiety, not only from the natural impulse of the mind which we have before noticed, to magnify the terrors of the "unseen," but also from a consideration of the fact that any difficulty in the way of the inspection is itself a proof that the construction of the fulcrum is of a nature to realize his worst expectations. The approximation to overhang the base, the ruggedness or irregularity of the declivity, circumstances on which its stability is principally dependent, are conditions in fact not only cognizable to the sight alone, but indicative by the facility with which they are submitted to its notice, of the actual state of the support itself in those particulars. The exclusion from his view may, in fact, be taken as the measure of the insecurity of the individual and the arbiter of his fears. In proportion as the fulcrum approaches a state in which actual peril must be incurred in the investigation, the mind becomes afflicted with the sentiments of its danger ; as soon as it has reached a point in which the precipitousness of its inclination has totally excluded it from the sight of the individual standing above, the stability of his position ceases to be altogether dependent upon its form, and becomes a question of consistency in the material of which it is constructed. With such a condition annexed, the fears of the individual assume a darker shade, and under the double influence of real and apprehended danger, amount to a paroxysm of agony which nothing but the certainty that the connection in question has no share in his support can either obliterate or appease. To that certainty the absolute knowledge that no such connection exists is alone sufficient. It is not enough that the continuity of the fulcrum be abstracted from his view; it must cease altogether to exist, and the mind must be aware of it, through the intervention of the senses. In short, it is not the *want of a visible connection*, but the *visible want of a connection* upon which the tranquillity of the mind is entirely dependent; a condition in which the aeronaut in his car is alone enabled

M

to participate. Relying entirely upon another quarter, he neither sees nor looks for a support, the insecurity of which he has reason to apprehend. The power by which he has been raised is all that he has to look to, and *that* unhesitatingly the mind admits to be all-sufficient for the purpose. Were but a pillar to connect him with the earth, or a rope to hang down, of sufficient magnitude to destroy these impressions by substituting a suspicion that *they* were the real means by which the equilibrium of the machine was maintained, giddiness and all the train of attendant symptoms would, I have no doubt, be the immediate conseqnence.*

As the aeronaut increases his distance from the earth, new circumstances arise to give birth to new relations, and call forth new sentiments of admiration and enjoyment. From regarding the altered aspect of the regions he has just quitted, his attention becomes forcibly directed to the condition and peculiarities of that into which he is now, for the first time, perhaps, about to intrude him-

* The preceding explanation of the absence of the usual symptoms of alarm and giddiness in the process of aerostation, was originally contained in a letter which I published in the Morning Herald, October 5th, 1836, on the occasion of a previous ascent, and from which (with some additions,) I have now extracted it. The letter, *minus* the explanation, which it has not been thought proper to repeat, I have been induced to give at length in a subsequent part of the present volume. See Appendix B, No. 1.

self. The clouds which he before beheld towering above his head, now begin to gather around and beneath him, and mingling with the various features of the scene, serve to diversify and adorn a prospect, whose chief characteristics are otherwise but sublime vacuity and unfurnished greatness.

With respect to the intervention of these bodies, however, the particular epochs at which they make their appearance, and the influence which they are capable of exerting upon the surrounding world,—it is impossible to affirm any thing with certainty. The circumstances upon which they depend and by which they are entirely modified—the influence of the weather, the condition of the atmosphere, the times and seasons of the year, the nature of the country, the very hours of the day, are matters too indeterminate to allow us to involve them in any general illustration of the career of the aerial voyager. Occasionally, for instance, clouds lie so low that, ere the balloon can be distinctly ascertained to have entirely quitted the earth, she has been received within their limits, and become entirely enveloped in their watery folds. Sometimes, on the other hand, these objects are disposed at such a height, that the balloon either never comes into contact with them at all, or if perchance she should have penetrated through one layer, continues to behold another, occupying a still remoter region of the skies above. At times again, these variable bodies are merely partial, affecting but a small portion of the aerial prospect, and arranged in different masses at different levels, or different stations upon the same level—a disposition I conceive the most favourable to the views of the aeronaut, as affording the best opportunity for that mingled display of earth and heaven which constitutes the chiefest source of his enjoyment; while, lastly, it will frequently occur that the whole face of the heavens is so completely overspread with clouds, that from the moment the aeronaut has once infringed upon their limits, until the actual conclusion of his career, earth and every thing that partakes of it becomes entirely excluded from his view. Of this nature was an ascent I once experienced, and of which I attempted to give an account in a letter published in the Times newspaper, October 21, 1836. To this letter, which I have been induced to repeat in a subsequent part of this volume,* I beg to refer the reader as containing the best illustration I am able to afford of the interference of these bodies, and of the particular effects and impressions to which they are calculated to give rise.

From the great variety of which they are susceptible, it is therefore pretty clear that very little can, even by the aeronaut himself, be affirmed with any degree of certainty as to the particular effects which the cloud creation

* See Appendix b, No. 2.

is likely to produce upon his voyage, before the actual moment of its execution. One piece of information, however, of rather a curious nature, a previous consideration of the state of the elements, under certain circumstances, enables him to deduce; I mean, with regard to the condition of the firmament above, at a time when, owing to its complete investiture with clouds, all view of that portion of the ætherial hemisphere is effectually suspended.

This information is founded upon observation, and is an inference from the state of the weather at the time with respect to the presence or absence of rain; as far as it goes it may be relied upon as perfectly established; to a degree of correctness indeed that few meteorological facts are capable of attaining. To reduce it to a general rule, therefore,—it may be asserted that, "whenever a fall of rain should happen to be present under circumstances like those detailed above, (namely, where the sky is entirely overcast with clouds,) there will be invariably found to exist another stratum of the same bodies at a certain elevation above the former;" and on the contrary, "whenever, with the same apparent condition of the sky, rain is altogether or generally absent, the aeronaut, upon traversing the canopy immediately above him, may infallibly calculate upon entering into an upper hemisphere, either perfectly cloudless, or so far destitute of such bodies as not much to interfere with the general character here bestowed upon it." This observation, which, independent of its value in other respects, is an addition to the stock of the meteorologist which he could never have obtained without the co-operation of the aeronaut, may be relied upon; it has been confirmed by the experience of Mr. Green, throughout a course of nearly two hundred and fifty ascents, and corroborated by that of various other aeronauts, both at home and abroad, with whom I have conversed upon the subject.* If the

* Two most remarkable instances confirmative of the truth of this observation occurred at the close of last year. On Wedneaday, the 12th of October, an ascent of the large balloon took place from the Vauxhall Gardens, under the circumstances comprised in the former illustration. The sky was completely overspread with clouds, and torrents of rain fell incessantly during the whole of the day. Upon quitting the earth, the balloon was almost immediately enveloped in the clouds, through which it continued to work its way upwards for a few seconds. Upon emerging at the other side of this dense canopy, a vacant space of some thousand feet in breadth intervened, above which lay another stratum of a similar form and observing a similar character. As the rain, however, still continued to pour from this second layer of clouds, to preserve the correctness of the observation, a third layer should by right have existed at a still further elevation; which accordingly proved to be the case. On the subsequent occasion of the ascent of the same balloon, the following Monday, (October the 17th,) an exactly similar condition of the atmosphere, with respect to clouds, prevailed; unaccompanied, however, with the slightest appearance of rain. No sooner had the balloon passed the layer of clouds immediately above the surface of the earth, than, as was anticipated, not a single cloud was to be found in the firmament beyond; an unbroken expanse of clear blue sky every where embracing

invariable co-existence of two circumstances can at all
be received as a proof of their relationship together, as
cause and effect, the share which the temperature has
in determining the condition of the clouds with respect to
the discharge of their aqueous contents, may be unequivocally inferred, and the above phenomena, upon such
grounds, easily explained.

To return from this digression: Varied as are the
positions of the clouds, and the forms under which they
present themselves, the station which they occupy in the
realms of space is confined enough, and, comparatively
speaking, but little removed above the immediate surface
of the earth itself. As a general rule, the natural region of
the clouds may be stated to be a stratum of the atmosphere,
lying between the level of the first thousand feet, and that
of one removed about ten thousand feet above it. Not but
that occasionally clouds may be found that trespass very
considerably on both sides of the bounds here assigned to
them; sometimes penetrating in wreaths of mist to the
depths of the lowest valleys, while, on the other hand, long
after the aeronaut has passed the upper level of these

the frothy plain that completely intercepted all view of the
world beneath. The close occurrence of these two cases, and
the very striking exposition they afforded, were in fact the
circumstances which first drew my attention towards the phenomena in question, and led to the adoption of the inference of a
mutual dependence between them.

fancied limits, some faint indications of their existence may still be seen, partially obscuring the dark blue vault above him; such excesses, however, are by no means frequent, and may, in fact, rather be considered in the light of exceptions to a rule than as evidences tending to impugn its general correctness.

It is certainly not to any inability in the medium itself to support them at higher elevations that is to be attributed this restriction of the presence of the cloud creation to the inferior regions of the sky; for where the aeronaut, with all his solid machinery and ponderous appurtenances, can penetrate and abide, assuredly there must be ample means of support for bodies which, by their unlimited powers of extension, can assume almost any degree of specific gravity, and, as it were, adapt themselves at command to media of almost every imaginable degree of tenuity. Rather to circumstances connected with their original formation,—the distance from the source from which they are drawn, the want of that degree of temperature necessary to determine their existence as vapour, perhaps also certain electrical conditions in the atmosphere affecting their dispositions to unite, in the form of rain — to these and other circumstances, unfavourable to their generation rather than to their support should, perhaps, be ascribed the confinement of clouds within such narrow limits, and the absence from the upper regions of the sky, of all those volatile bodies, which we, in respect of our own more humble stations, are wont to consider as the emblems of aetherial pre-eminence and the types of all that is remote, lofty, and sublime.

The simple circumstance of their comparative elevation, however, is capable of exerting but little influence upon the prospect of the aerial voyager, unless indeed he is contented to confine himself to the mere threshold of the element he proposes to survey; his increasing altitude very soon places him in a situation from whence all things appear equally depressed, and from which indeed he could with difficulty ascertain, by the mere aid of his sight, whether the clouds he is observing are really reposing upon the surface of the earth or seated at an elevation of several thousand feet above it.

Should the condition of the sky now prove to be of the nature of that alluded to,—where, for instance, a dense layer of clouds completely intercepts all view of the earth, the aeronaut will probably have an opportunity of observing another phenomenon connected with the disposition of the vapoury strata,—the beautiful manner in which, even when under the influence of rapid motion, they seem to accommodate themselves to all the variations of form in the surface of the subjacent soil, rising with its prominences and sinking with its depressions; displaying, in short, a " counterfeit presentment" of the country over which they lay, and enabling the spectator to form, as it were, a sort of phrenological estimate of the character and disposition of the material world within. Indeed, I have heard Mr. Green declare that, with the bird's-eye knowledge of the country his long experience has conferred upon him, he has frequently been able to determine before hand the district into which he was about to descend, at times, when from the general concealment of the landscape, such information must have been otherwise altogether unattainable.

The most favourable arrangement, however, for the views of the aeronaut who feels an interest and a gratification in the study of the picturesque, is decidedly that in which the clouds, from their broken and disconnected nature, spread at unequal intervals throughout the empty space of air, admit occasional glimpses of the earth in different directions, and passing gradually over its surface, in succession reveal an ever-varying prospect, to the constitution of which heaven and earth so equally contribute that it is difficult to determine to which to award the palm. Such scenes, however, are not for the pen, scarcely even for the pencil : for who by signs can hope to justify a prospect which is much less dependent for its effect upon the materials of which it is composed than the manner in which they are examined—upon its own attractions than sentiments pre-existing in the mind of the person by whom they are enjoyed ?

But see ! the balloon has already passed the limits we have assigned to these " hoary riders of the blast," and is

now rapidly pursuing her course into realms hitherto unknown to man, even on the summits of the highest mountains accessible to his exertions.   Here then let us pause
for a moment to take a hasty glance at the nature and
condition of the scene around, and the sentiments and
impressions it is naturally calculated to produce upon the
mind of the aerial beholder.

With less numerous subjects for the exercise of his
senses, it must not be supposed that these, the remoter
districts of the etherial domain, are by any means deficient
in grounds for enjoyment even of the very highest order.
It is true, here are none of the usual combinations of form
and colour which give such zest and variety to the terrestrial landscape; none of those delightful sounds which,
pervading the whole habitable world, maintain the idea of
animation even in the veriest desert; none of those fra-
grant exhalations by which— as it were, the music of the
vegetable world—every tree and flower, gives vent to its
own particular sentiments. These, it is true, there are none
of; but even, in their very absence, the aeronaut finds a
source of gratification, more intense at any rate, if not more
interesting, than any with which their presence could have
supplied him. Undisturbed by the interference of ordinary
impressions, his mind more readily admits the influence of
those sublime ideas of extension and space which, in virtue of his exalted station, he is supremely and solely calculated to enjoy. Looking out from his lofty car, in every

direction save one, and *that*, one from which similar sentiments never before proceeded, a boundless blank encounters his gaze, unbroken, except, perhaps, by bodies whose
thin aerial forms and fleeting aspect constitute them sole
fitting occupants of such domains. Above and all around
him extends a firmament dyed in purple of the intensest
hue, and, from the apparent regularity of the horizontal
plane on which it rests, bearing the resemblance of a
large inverted bowl of dark blue porcelain, standing
upon a rich mosaic floor or tesselated pavement. In
the zenith of this mighty hemisphere—floating in solitary magnificence—unconnected with the material world
by any visible tie—alone—and to all appearance motionless, hangs the buoyant mass by which he is upheld.
The world he has quitted, and that towards which he
tends, seem to his fancy, almost equally remote; and as he
endeavours to scan the empty vault that divides him from
the earth, he involuntarily imbibes a sentiment of immense vacuity, which no other situation and no other
scene is capable of communicating. It is not that the
interval through which his eye has to travel in reaching
the ultimate scope of its views, is really so vast; for what
after all are the few thousands that constitute the utmost
elevation of the aeronaut, compared with the countless
myriads that separate him from the nearest visible object
of the external universe, and which, stretching for ever
above his head, lie ready at all times to meet his eye

whenever he pleases to direct it thither? It is not therefore in the mere amount of intervening space itself, that
consists the peculiar force of his impressions, but that,
bounded to a certain extent by known and recognized
limits, in the effects produced upon them by distance, he
has a measure for its magnitude to which the mind is enabled to refer. From such a resource he is entirely precluded who seeks to fathom with his eye the boundless
abyss of infinite extension; no appreciable object there
appears to intercept his view or regulate his judgment;
he sees nothing, and seeing nothing can assuredly form
no definite conception of how much it is capable of including. In short, to form an estimate of space from
observations directed towards the realms of infinity,
requires an *active* exertion of the intellect of which all
people perhaps are not susceptible; whereas to the individual who studies it thus, as it were, measured off from
the mass, the impression suggests *itself*; the mind is *passive;* the idea is presented to it, and will not be refused.
As to the comparative amounts, they signify but little to
the general effect; beyond a certain quantity the mind
is incapable of containing, even if the eye were capable
of conveying an idea of extension. To the human judgment thus restricted, the quantity which divides the earth
from the aeronaut at his greatest elevation, inasmuch as it
is comprehensible, is far more effective than the utmost ex-

tent of infinity to which his eye could penetrate, without the aid of such expedients; not from the reasons here detailed, alone; but likewise because, taken in an unwonted direction, and one where its occurrence is generally coupled with notions of insecurity and fear, both novelty and awe combine to give a zest to the sentiment, from which the extension of his view into another quarter is entirely exempted.

A striking illustration of the influence of matter in determining the mind to admit the full force of these impressions, is afforded in the contemplation of a solid body in the act of falling from the car, while at a superior elevation, and tracing with the eye its progress as it descends towards the earth;—the silent magnificence of the abyss into which it plunges, the complete isolation of the beholder, the apparent infirmity of the fragile vehicle over the side of which he peers with impunity; then the sudden force with which the body appears to escape from his hand, as if violently launched from a machine, and the equally sudden retardation which, after it has dropped a few feet, it seems to have experienced, together with the length of time it afterwards remains in sight, and the comparative slowness of the changes that increasing distance operates in its dimensions—all natural consequences of the event under the peculiar circumstances of the case, which no less by the sympathies they involve than by the indications they afford, awaken a mingled sentiment of sublimity and space, no where else, and by no other means, to the like extent, acquirable.

Of these phenomena by which the fall of a body from the balloon is attended, only two require comment;—the apparent retardation of its progress, following upon so rapid a commencement;—and the length of time, which, in despite of the onward course of the balloon, it continues to be discernible in the same direction. The former of these is an impression analogous to that by which the ascent of the balloon itself, *in limine*, is accompanied, and of which an explanation has already been given in a previous part of this appendix.* It is unnecessary to do more here than to remind the reader that the effect upon the eye is precisely the same, whether the spectator be himself removed from the vicinity of the object, as in the former instance, or the object be removed from the eye of the spectator, as in that at present under consideration; the impression of unusual rapidity, displayed in the first fall of the body from the car, being no doubt frequently enhanced by the occurrence, at the same time, of a similar motion in a contrary direction on the part of the balloon from which it is dismissed. The other phenomenon referred to—namely, the long-continued presence of the falling body in the same direction, notwithstanding the onward progress of the balloon,—is

* See page 138.

founded upon such very simple rules that, to the scientific reader, no explanation is requisite. For the advantage of others, however, it may be as well to observe that, by the immutable laws of matter, motion once communicated to an inanimate body must ever continue to influence its progress in the original direction conferred upon it, until it has encountered some other substance to which to impart it. Impressed, therefore, with the motion of the balloon at the time, every thing that quits the car, without a special impulse in another direction, must continue perpendicularly beneath it, until it reach the earth, or mayhap encounter in its descent some current of air proceeding from a different quarter, in the resistance occasioned by which, its original motion becomes gradually dissipated and destroyed.

In the midst of this immense vacuity, which, with feeble pen, we have vainly endeavoured to depict, it is not to be wondered should a sense of solitude, to a degree never before experienced, form the predominating character of the feelings with which the aeronaut is, as it were, forcibly impelled to regard the scene around him. Utterly abstracted from all contact and communion with the habitable world, environed and upheld by an invisible medium, without a single object to interrupt the drear monotony of all about him; nothing can be more perfect than the state of isolation in which he is placed; and as he looks out from his airy domicile upon the immense void that every where surrounds him, and regards the exiguous spot he occupies in its vast enclosure, he is driven to acknowledge the force of new impressions, and for the first time in his life *is* really, and *feels* alone. To the production of these sensations, no other situation is at all competent. Likest to it, though still far removed from the complete enjoyment of its conditions, is that of a boat at sea; but *boats* never are at sea, in the full sense of the word, and ships, besides the scene of animation which they invariably and inevitably display, have far too much of man to permit the existence of a feeling which diminishes, in fact, only in proportion as it is participated.

But the most powerful auxiliary to the sense of solitude peculiar to the situation of the aeronaut, is the extraordinary silence that qualifies the region of his new adventure. No words can, in truth, sufficiently represent the remarkable condition of the skies with regard to the absence of sound, or convey any just notion of the extent to which that particular condition is capable of affecting the human organs. Indeed, to describe a state of things depending upon the *negation* of a cause in such a manner as to produce the idea of a *positive* effect, is at no time an easy task; when, however, to the ordinary difficulties arising from the defects of language, is added the want of a proper community of sentiment between the parties, little can be expected to be comprehended by a mere perusal of phenomena however accurately detailed, and however correctly accounted for. All, in fact, that can be said

N

upon the subject by way of illustration is, that here, and here only, absolute silence can be said to have any existence at all; in every other situation in life, on the summit of the highest mountain, in the depth of the lowest cavern, in the desert and on the sea, in the dead of night, and the stern repose of the veriest calm, sound, to a certain extent, (however it may escape our habituated faculties,) always exists. The vicinity of solid bodies, the resistance of the air, the influence of the changes of temperature upon adjacent matter, nay, the very process of vegetation itself are so many sources of sound from which man is never entirely free until the aerial car has snatched him from their influence, and transported him to regions where none such are ever to be found.

Yet is not the unwonted absence of the actual causes of sound, the only peculiarity under which the faculty of hearing is exercised in the upper regions of the atmosphere; for perhaps at no time is the attention of the aeronaut so forcibly impressed with the singularity of his situation in that respect as when the natural tranquillity of the surrounding medium is under the temporary influence of disturbance from artificial causes. The contiguity of solid matter has in fact another task to perform than the mere generation of sound, in the modifications to which it is incessantly subjecting it, during every stage of its continuance. Scarcely has a sound been promulgated in ordinary situations than it is immediately encountered by a thousand obstacles that alter, impede, protract, derange, and qualify its vibrations, and, by the manner in which they interfere with their simplicity, produce in the same effect upon their impressions as the intervention of the obscurating medium, already described,* upon the objects of the sight; confusing their outlines and depriving them of that sharpness of contour, and vividness of character which in fact may be said to be their natural or at least their legitimate condition. It is true that of such modifications in ordinary circumstances the ear takes no note : unconscious of the effects of sound in its pure and simple state, it suffers no particular impression from the presence of a condition to which it is habituated, and from which it has never at any time been absolutely free ; nor is it until it has been transferred to a situation where these modifications no longer exist, that it becomes aware of their influence, and able to appreciate their absence. Such is the advantage which it enjoys in the balloon, and such the restrictions under which the sense of hearing is exercised in the upper regions of the atmosphere. There—situated apart from all contact or intercourse with the solid world—no sound ever reaches the ear more than once, or continues beyond the natural duration of its own primary vibrations. Deprived

* See page 145.

in a measure of all those artificial asperities, by which it is usually distinguished, its character becomes totally altered, and like the landscape to which we have before figuratively referred, it strikes upon the senses in all its native purity, sharply, simply, strongly, and perspicuously delineated. With such qualifications, the casual occurrence of sound is consequently attended with even more uncommon effects than the natural stillness of the surrounding medium, extraordinary as that may be. The various interruptions it occasionally receives from below, the barking of dogs, the lowing of cattle, the tinkling of the sheep-bell, the exercise of the different instruments of the artificer, the saw, the hammer, and the flail, when at a moderate elevation—the shot of the sportsman, the reiterated percussion of the fulling and other mills, the discharge of artillery, and the voices of those beside him, at a greater distance from the earth, are all sources of interest to the aeronaut, which please no less by the associations they awaken than by the peculiar effects with which they are made apparent. Of all the sounds, however, which meet the ear of the adventurer in these exalted regions, none appears to me to bear with it so impressive a character, or to be productive of such awful sensations as that occasioned by the snapping of the valve in the upper part of the balloon, when in the act of closing after some occasional discharge of gas in the course of the excursion.

The sudden sharpness which it displays in common with the rest, the unusual direction from which it issues—a direction from whence no sound is naturally expected to proceed, the intimate connexion between it, the office it has to perform, and the fate of the aeronaut by whom it is worked, together with the drum-like intonation which instantly supervenes, caused by the extreme tension of the silken dome, in the cupola of which it is situated, and which in a manner serves like a sounding-board to sustain and prolong its fleeting impulses,—all combine to bestow upon it an effect and a sentiment which belong to no other sound, and are experienced in no other situation.

To the enhancement of all these effects, as well indeed as of those perceivable in the exercise of all the other senses at immense elevations, the rarefaction of the air, and the temperature of the region no doubt likewise essentially contribute: not by increasing the ability of the medium for the conveyance of the impression, (for in respect of sound, and perhaps of all but sight, such conditions are rather detrimental to its qualification for such a purpose,) but by their action upon the organs themselves, producing as it were a slightly morbid state, which renders them more susceptible of the impression ; as we occasionally perceive to be the case in some diseases, where the irritability of the nervous system attached to some particular organ, becomes so much increased, that circumstances which otherwise would have passed unnoticed, produce, not only powerful, but sometimes even painful excitations of the sense to which they are directed. In the exercise of the faculties of hearing, seeing, and smelling, these advantages are most strongly and strikingly experienced, especially the latter, owing as well to the assistance which the rarefaction of the air gives to the dispersion of the volatile particles of the odoriferous essence, as to the morbid adaptation it confers upon the organ to receive them.

With the increase of his elevation, of course, keep pace all those phenomena which depend for their effect upon the diminished density of the surrounding atmosphere; and as the aeronaut draws nigh to the highest point ascribed to the ordinary course of such adventures, begin to exhibit, in the altered characters they have assumed, proofs of the unwonted circumstances under which they are experienced. In nothing is this more strongly exemplified than in the appearance of the firmament itself. The colour of the sky, the increasing darkness of which we have already noticed, has now reached a pitch of intensity so great as scarcely to seem compatible with the ætherial consistency of a mere gaseous accumulation, and almost to warrant the impression of an approach to something whose limits are more substantial and defined. M. Gay Lussac, in his second ascent from Paris, in which

he attained an excessive elevation,* represents the colour of the sky, especially about the zenith, as observed from the highest point in his excursion, to be on a par with the deepest shade of Prussian blue; an observation, the justness of which will be felt and acknowledged by all who have ever proceeded to any considerable distance from the surface of the earth.

In order more properly to comprehend the ground upon which these alterations are chargeable, it is absolutely necessary that some insight should be obtained

* The greatest altitude to which any balloon has ever been *known* to ascend, is that accomplished by M. Gay Lussac this voyage, and is calculated at seven thousand and sixtee French metres, or twenty-two thousand nine hundred a seventy-seven feet, four inches, above the level of the sea. F an account of this voyage, which is interesting on more accou than one, see the memoir attached to his name in the A pendix c, in a subsequent part of this volume.

I am aware that other aeronauts lay claim to higher honou M. Blanchard, for instance, states himself, upon one occasi to have attained an elevation of nearly thirty-two thousand —an assertion upon which MM. Margat, Garnerin, Robert and others, his successors in the trade of aerostation, have, f time to time, considerably improved. For these pretensi however, there is not the slightest foundation; nor would a matter of much difficulty to demonstrate, that the ball they employed, (with the dimensions of which we are well quainted,) could not, even if inflated with the purest hydro have supported their simple weights at much above one-hal elevation they would fain have persuaded us they attained.

into the nature of the appearance of the firmament in general, of which the phenomena in question are merely modifications.

That the aspect which the heavens present whenever circumstances permit us to enjoy an unobstructed view of them, whether it be the azure complexion of the glowing day, the sable livery of night, or the milder shades of twilight grey, by which the transitions from the one to the other are invariably distinguished, is not a condition to that effect in any way inherent in the body of the atmosphere itself, a variety of arguments satisfactorily enable us to determine. The most striking of these are deduced from the following observations; first, that whenever any portion of it happens to be separated from the general mass above it by the intervention of clouds, no symptoms of the prevailing tints can ever be detected in the part so intercepted, although by reason of its superior density, it may, in respect of actual quantity, equal, if not exceed, all the rest of the medium which lies beyond it; and secondly, that instead of appearing lighter the less the quantity through which the eye has to penetrate, (as in the case of all other known transparent bodies, which possess a colour of their own,) its shade continually increases in intensity the more the superincumbent mass is diminished by the translation of the spectator to a higher position within it.

That the existing appearance is not, as some

have vainly imagined, a general contribution raised by reflexion from the surface of the earth, and modified according to the particular predominance of the solar influence, is likewise a conclusion which with equal clearness we collect, partly from a consideration of the fact before observed, that when the communication happens to be dissolved by the interposition of clouds, the character in question, whatever it may happen to be at the time, entirely disappears in the portions contiguous to the earth, while neither in quality or intensity is the slightest alteration perceptible in those which lie beyond; and, partly from the observation, that, in all countries, under every variation of climate and through every change of season, the genuine aspect of the sky is virtually the same.

Since then the appearance of the heavenly arch is neither a quality which resides in the substance of the atmospheric volume, (the only material obstacle of whose presence in that direction we are aware,) nor is obtained by the process of reflexion from any thing which exists on this side of the space which it subtends, it is evident that no other way remains by which it can be accounted for than by a reference to the condition and modification of something which occupies or proceeds from the interval which lies beyond. To enable us to justify our conclusions upon this score, we must first endeavour to ascertain what is the natural aspect of boundless space, or what would be the appearance of the mighty vault of heaven could we but direct our gaze into its vast inclosure, unencumbered by the presence of a medium of refraction. The inquiry is fraught with awe no less than interest. It almost seems like intruding upon the especial domains of the Almighty to attempt to tear the veil from the charms of boundless space, and expose the secrets of a condition of which our visual faculties but render to our senses an account as mysterious and imperfect as that which our mental ones with their utmost exertions are able to convey to our understandings. To say that the heavens, or that portion of space beyond the limits of our atmosphere, which we are wont to honour by such a name, possess a colour, would be, in truth, to employ a misnomer. Space, infinite space, unfilled with matter, must really be devoid of colour; and excepting in the bodies it contains, must ever present to the eye of him who views it in all its natural majesty the terrific aspect of a black and fathomless abyss. To confer the idea of a colour, or permit the rays of light in any way to vary its appearance, the presence of a transparent medium is absolutely requisite. Near the body of the earth, and of most of those other objects with which astronomy has made us acquainted, such a medium obtains in the atmosphere by which they are surrounded; and in the effects of this atmosphere upon the aspect of the black vault behind, lies the only condition that exists to vary the visual presentment of void and infinite space. By the combined exertion of two of its properties, this result is accomplished ; first, by the diffusion of the white rays of light, whereby the extreme obscurity of the background is tempered into paleness,* and in the next place by the occasional interposition of a new colour obtained from the rays of light transmitted through it from above, whereby the original in its subdued intensity becomes at times invested with a colour compounded of them both.

To the full force of the former influence, much is no doubt contributed by the presence and disposition of solid matter in the neighbourhood of the field of view, by means of which the adjacent portions of the surrounding medium become as it were charged with the superfluous rays of light from various quarters, tending not only to distract the eye of the spectator and to confuse his prospect, but also to reduce by the copious admixture of white light the natural intensity of whatever object may happen to be exposed to it. To what an extent the diffu-

* The influence of atmospheric irradiation upon the aspect of the void space by which it is backed, was first suggested by the father of the pictorial art, the celebrated Leonardo da Vinci, and afterwards revived and adopted by M. De la Hire, as a probable cause of the azure colour of the sky. In support of this theory, a variety of experiments were adduced to prove that black, when beheld through a white or colourless medium, always inclines to assume a blue or azure tint. The truth of

sion of light, so supported, is competent to produce the results ascribed to it, we see clearly evinced in the extraordinary effects occasioned by the temporary suspension of its influence. No sooner has any interruption taken place, either through the discontinuance of the supply, its artificial exclusion from the field of view, or a diminution in the capacity of the medium for its conveyance, than the eye immediately reads the change in the unwonted darkness of the sky, and (when promoted to a sufficient extent,) in the renewed appearance of the luminous bodies which it enfolds. Of this, illustrations are naturally afforded in the approach and presence of night,—during the continuance of a solar eclipse, when the obscuration has reached a sufficient magnitude,—or still more remarkably upon the occasion of any unwonted rarefaction in the atmosphere, such as is frequently found to

these deductions, however, is more than doubtful. The proper product of black and white, or in other words, of *no* colour with a combination of *all* colours in the proportion in which they exist in solar light, is always grey : nor do I believe that any other colour could ever result from their admixture, no matter the proportions or the means whereby it might be sought to unite them. To what to attribute the fallacy of their conclusions I really know not ; unless indeed it might be owing to an incorrect estimate of the colours of the substances employed in their experiments, or the want of proper precautions to avoid the introduction of colours obtained by refraction from the transparent media through which they were examined.

precede a sudden change from fine to foul weather; and
artificially, and with equal effect, by removing to a distance from the surface of the earth in a balloon, or by
proceeding in the other direction to the bottom of a well
or vertical shaft, sufficiently deep to afford a complete
protection against the influence of the circumambient
irradiation.

But the mere diffusion of light, to whatever extent it
might be carried, although it might alleviate the intense-
ness of colour in any object, could never avail to give it
a new one, or make that assume "the front of azure
blue," whose legitimate aspect was unmitigated sable.
This is a result which requires the intervention of another
property in the medium; such a one, for instance, as that
alluded to, whereby the rays of light transmitted through
it from above, are made to affect a colour suitable to the
compound required.* As the intensity of this colour, as

---

* With the existence of such a property we were first made
acquainted by the researches of Sir Isaac Newton; who, having
ascertained that vapours, when about to condense and coalesce
into drops, first become of such a size as to elicit the blue rays
of transmitted light, was induced to attribute the azure colour
of the sky to a condition particularly favourable to the exercise
of such a property, which it was presumed existed only in the
remoter regions of the upper air. The existence of a vapour at
all times present in the atmosphere, a circumstance essential to
the views of Newton, was, however, a weak point in his theory,
which has induced subsequent inquirers to look for some more

well as that of the black vault by which it is supported, is a quality subordinate to the influence of atmospheric illumination, whatever tends to the abatement of that illu-

permanent quality in the same quarter upon which to charge the occurrence of the observed phenomenon. Accordingly, after a variety of experiments, a French philosopher, M. Bouguer, considered that he had solved the difficulty by referring the separation of the rays in question to a difference in the *momenta* of the different constituents of solar light, whereby the red alone, supposed to be possessed of superior motive energy, made their way unobstructed to the surface of the earth, while the blue, considered of weaker impetus, unable to advance, remained behind to imbue with their particular colour the remoter strata of the atmospheric fluid by which they had been absorbed. These views of M. Bouguer, sufficiently ingenious considering the then state of the science, the recent establishment of the theory of undulations requires us to interpret after another form. Admitting the exclusive progress of certain rays, but rejecting the grounds of different momenta by which it was formerly wont to be explained, reference must now be had to another principle, namely, the *critical angle of incidence*, whereby the blue rays, instead of entering the body of the atmosphere are reflected at an angle, and would be altogether dismissed unnoticed, but that owing no doubt to the extreme tenuity of the upper strata of the atmosphere, they have already proceeded to a considerable distance ere they have encountered sufficient consistency to determine their return. For the benefit of the unlearned, however, we may as well observe that it matters nought in the least to the subject in hand, which, or whether any of the views here proposed, be the correct one. It is enough for us that there is a property of the nature referred to, existing in the upper strata of the atmosphere; and *that* is

mination, either by the curtailment of the supply, its artificial exclusion from the field of view, the diminution

a fact of which we have sufficient proof in the evidence of our senses.

Indeed but that the limits of a note are too restricted for the purpose, it would not be a difficult matter to point out occurrences which do not appear to consist with any of the views here taken of the subject. For instance I do not see upon which of these grounds can be explained the phenomenon, (very frequently observable upon occasion of the setting sun,) of the complete determination of the blue rays to the quarter directly opposite the seat of that luminary, leaving the rest of the heavenly hemisphere comparatively devoid of any such inclination. In all these cases, the blue, if really obtained by the decomposition of solar light *in transitu*, must not only have traversed one radius of the atmospheric horizon in company with the red, but afterwards exclusively continued its course to the further extremity of the opposite one. Another circumstance, apparently incompatible with the foregoing views, is the extraordinary blueness discoverable upon the occasion of a sudden rarefaction in the atmosphere : were the blue, in these cases, merely the complement of the red, previously interrupted in its passage, its subsequent intervention should only have restored the whole to its primitive condition of a colourless compound.

These objections, I beg to observe, are by no means intended to impugn the correctness of the theory of undulations as a system explanatory of the nature and properties of light in general. On the contrary, it is upon the assumption of its superiority, that we are led to question the accuracy of any views to which its principles seem in the slightest degree irreconcileable.

of the capacity of the medium for its conveyance, or the remotion from a neighbourhood where its natural amount is increased by adventitious reflexion, tends likewise to increase the intensity of the sky, and bring out more forcibly the natural obscurity of the ætherial scene. Of these latter, the ascent in the balloon is a striking illustration. Diminishing at once the density of the medium, and the amount of its terrestrial irradiation, at every step he recedes from the surface of the earth, the aeronaut obtains in the darkened aspect of the heavenly arch unerring tokens of his approach to the nether limits of the void and infinite gulf that lies beyond him ; and, I have no doubt, could he but continue his course until he had attained the outward margin of the atmosphere, he would, upon directing his view into the realms of vacuity, behold an impenetrable abyss of perfect blackness, in which every visible source of light would stand like a disk of solid flame, unaffected by the vicissitudes that, for one half the period of their revolutions, exclude them from the eye of the terrestrial spectator.

How long before that extreme was attained, the latter part of this description would have been realized, and the heavenly bodies revealed to the naked eye in broad daylight, I cannot take upon me to determine ; if, however, the obscuration of the sky, (upon which the occurrence of the phenomenon in question entirely depends,) were to continue to increase at the same rate we observe it in the earliest stages of the ascent, (and there is every reason to admit the conclusion,) I do not think that the possibility of witnessing such an occurrence is entirely beyond the hopes of the aeronaut adventurous enough to attempt it, and provided with means corresponding to the peculiar exigencies of the occasion. Some indeed there are, who, even without these advantages, pretend to have attained situations in the ordinary exercise of the art, from whence the existence of such a phenomenon could clearly be discerned : I should rather, however, suppose that this assertion was merely an exaggeration of the fact, that at their utmost altitude they were able to distinguish the presence of the heavenly bodies sooner than they could have been perceived by persons situated upon the actual surface of the earth ; an assertion which in fact amounts to nothing more than what we know would have been experienced under any circumstances of superior elevation, no matter how slight. With every degree of removal from the plane of the terrestrial horizon, the capacity of the surrounding medium for the diffusion of light becomes diminished, and the contrast in favour of the spectator, (which is the only cause of their suppression,) being weakened, the stars of course make their appearance at an earlier hour than they would if that contrast had to wait the decline of day to effectuate its abatement. The question is evidently, therefore, one of comparison, and is as easily put to the test by ascending to the top of a hill as by encroaching upon the limits of the sky in a balloon. If I mistake not, something of the kind is mentioned by the elder Saussure, in his account of the first ascent of Mont Blanc, as having been observed upon the summit of that mountain; an observation which has been repeated in other parts of the globe by all travellers who have ever succeeded in attaining great elevations upon the surface of the earth. That such a result could be produced by an artificial exclusion of the light, as for instance in the bottom of a deep well, or any other excavation of sufficient profundity, was a fact well known to the ancients, who, in Egypt especially, were in the habit of constructing pits on purpose to aid them in their study of the heavenly bodies; many relics of these subterraneous observatories remaining to the present day to bear testimony to the industry and acquirements of those learned Pagans.

In consequence of the increased removal from the vicinity of the earth, the temperature of the surrounding medium has become considerably reduced, and were it not for the absence of all atmospheric motion would, no doubt, be severely manifested to the feelings of the aeronaut. At what particular period of the ascent, this decrease attains a minimum, or indeed whether such a result exists within the range of aeronautical adventure, I am not able, with any degree of certainty to state. The solution of the question, which is undoubtedly an interesting one, depends chiefly upon the point to which the calorific influence of the earth's radiation extends, and is only to be arrived at by a long-continued series of experiments and observations. Of course, the results here as elsewhere will be found to vary with the climate, the season of the year, the hour of the day, and the state of the atmosphere at the time prevailing. In one respect particularly, the latter is capable of exercising a very sensible influence over the thermometrical condition of the upper regions; I mean where clouds to any amount intervene, whereby a large extent of reflecting surface becomes presented, and a very considerable portion of the heat of the solar rays returned into the body of the atmosphere which lies above. As this is an arrangement of the sky more frequently to be met with in winter than in summer, it follows singularly enough that the effects of a low temperature are much less likely to prove injurious to the aeronaut in the exercise of his art, during the more rigorous portions of the year, than those which every where else come under the denomination of the milder and more serene.

Considering, therefore, the number and irregularity of these disturbing causes, it will appear pretty evident that no exact measure of the temperature, and consequently, no just representation of its effects upon the human frame could be afforded that would apply with equal correctness to all the circumstances under which it might be tried. In general, however, where there are no clouds to interfere with the natural progress of the sun's rays, a temperature of thirty-two degrees of Fahrenheit may be expected to be encountered at an elevation of about seven or eight thousand feet above the level of the sea—that is to say, in these climates; the region of eternal frost, or as it is geographically termed, the *line of perpetual snow*, entirely depending upon the latitude of the place, and diminishing in elevation in proportion as it increases its distance from the terrestrial equator. Beyond this altitude, the temperature, as before, keeps continually decreasing, though with waning rapidity, and at an elevation of twenty-two thousand nine hundred feet, the thermometer as observed by M. Gay Lussac, had fallen to nine degrees five-tenths of the centigrade division, or seventeen degrees one-tenth below the freezing point of water according to our usual method of computation.

Supposing, however, the state of the temperature to have been in any instance even twice as low as that above indicated, still there is much reason to question whether at any time the sufferings of those exposed to it can have been so severe as many would fain incline us to believe. Certain allowances ought, no doubt, to be made for the constitutional peculiarities of different individuals; and much ambiguity must always be expected to prevail where personal feeling is the subject of discussion, and the sense itself the only test to which it can be subjected. But with all these admissions, there is still sufficient evidence in the experience of those who both naturally and by artificial means in the way of experiments have placed themselves in circumstances of like exposure upon the earth, to authorize a doubt that much inconvenience ever did or could accrue to the aeronaut, who in the exercise of his vocation may have penetrated to the utmost limits his means in other respects would allow him.

Having now attained the highest point to which it is our intention at present to proceed, we will pause for an instant to take a cursory glance at the earth, ere we prepare to incline our journey thitherward again. The landscape which, for some time back, has been gradually displaying symptoms of decreasing perspicuousness, has now suffered so much from the effects of distance that it is not without difficulty that any of its ordinary features can be distinguished. Not that any abatement appears to have taken place in that vividness of contour which we have before observed to be the never-failing peculiarity of the terrestrial scene when viewed from the car of the balloon; but that the objects themselves have now become so much reduced in size, that many of those, the most familiar and characteristic, have altogether become extinct, and the rest so much estranged in their appearance as to contribute but little to the recognition of the prospect of which they form a part.

Amid this scene of universal disfiguration, all perception of comparative altitudes is utterly out of the question. Removed to such a distance from the eye, and solely submitted to a vertical examination, the whole face of nature, in fact, appears to have undergone a process of general equalization; the houses and the trees, the mountains and the very clouds by which they are capped, have long since been consigned to the one level; all the natural irregularities of its surface completely obliterated, and the character of the *model* entirely superseded by that of the *plan*.

It has frequently been inquired of me, whether under circumstances of such excessive elevation, any symptoms of convexity can be detected in the appearance of the horizontal plane, such as a knowledge of the real form of the terrestrial globe might have authorized us to expect. When, however, we consider the immense disproportion which exists between the actual diameter of the earth, and the utmost altitude to which man ever did or could attain above its surface, we shall cease to look for such a result, or be surprised at observing the deficiency. Were we to assume an elevation of forty-two thousand feet, (which is nearly double what has hitherto been accomplished,) as the *ne plus ultra* of aeronautical enterprise, still, computing the earth's radius at four thousand miles, and reckoning five thousand two hundred and eighty feet to each mile, the prominence of the spectator beyond the surface of his horizon would even then amount but to the thousandth part of its extreme lateral extension: in other words, he would have only reached a distance beyond the plane of his vision, as great as the thickness of the smallest letter we are now employing, (estimated at the hundredth part of an inch,) would project upon the face of a globe of ten inches in diameter. In short, his newly-acquired position would no more enable him to discern the sphericity of the earth, than the eye of a beetle would convey to it an idea of the convexity of the mountain whose rounded summit it was slowly labouring to ascend.

In answer to this, the reader may perhaps suggest the well-known phenomenon of a ship at sea approaching from a distance, and adduce the gradual disclosure of its parts as an evidence of the possibility of obtaining, under a favourable conjuncture of circumstances, ocular testimony of the nature which we have here attempted to disprove. The example, however, is by no means a case in point. It is not the *sphericity* of the earth that the eye in such cases observes, but merely its *effects*; and therein can no more be considered as reading the convexity of the earth, than a man looking at his shadow upon the wall can be said to be observing the taper which stands upon the table at his back. Without the convexity in question, it is true the phenomenon observed could not have taken place; but neither, on the other hand, would the convexity in question have been observed, had not the said phenomenon been present to disclose it.

But even if the conclusion were otherwise, still the

cases are by no means analogous, nor could any argument be drawn from the capacity of the eye in the one instance, to sanction the expectation of a similar result in the other. Comparative altitude, which is, in fact, the only test of prominence, is a condition the knowledge of which is only acquirable by means of an examination conducted at right angles to the plane of extancy; or, in other words, by observing the *profile*, more or less, as it appears represented upon the substance or substances which may happen to be aggregated in the rear. From the enjoyment of this advantage, the aeronaut by his position is thoroughly precluded; all his views are necessarily downward, and all his perceptions of form confined to the observation of surfaces projected upon the plane beneath him.

With all these considerations, however, the inexperienced reader will, no doubt, learn with surprise that the real form of the earth, as beheld from the car of a balloon sufficiently elevated in the air, is absolutely the very reverse of that which a first view of the case may have hastily inclined him to expect. Such, however, is undoubtedly the fact. So far from following the course dictated by the true conformation of the earth, and sinking in proportion as they recede, the edges of the terrestrial plane actually assume a contrary inclination, and, rising as the aeronaut increases his altitude, realize in their progress the appearance of a vast bowl or basin extended on all sides around him.

Unexpected as this phenomenon may at first sight appear, it is, nevertheless, but the natural consequence of the laws of refraction acting under the peculiar circumstances of the case. Diverted from the straight course which the sight would at all times pursue, were it unobstructed by a medium of refraction, the lines under which the various objects are beheld, become gradually inclined upwards, referring the objects themselves to points in their new positions, at distances from the eye of the spectator equal to those at which they are actually situated. This will be better understood by a reference to the annexed

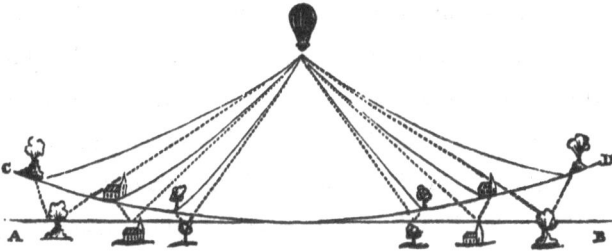

diagram, in which the station occupied by the aeronaut is represented by the small balloon ;—the direction which the sight would have travelled, had there been no refracting medium, by the dotted lines, and that which, in consequence, it is forced to assume, by the plain ones. As the *distance* is not falsified by the refraction, the various objects upon the terrestrial horizon A B, (as there depicted,) will, in appearance, be transferred to stations equally remote from the eye, and be found occupying a curve, C D, formed by a close continuation of points in the refracted lines of vision, equidistant from the eye with those which they represent upon the horizontal surface of the earth.

But it is now time to conclude. Too long already I fear have I detained the young adventurer in the realms of upper air; more especially as this is his first attempt, and he must no doubt feel anxious to return and quell the fears of his family and friends below. We will, therefore, pull the valve, and commence our descent.

And let not the reader suppose that in this seemingly simple phrase consists all that is required to the achievement of this most important operation; and that the aeronaut has nothing to do, when he desires to terminate his excursion, but to pull the valve, and take his chance for the result. It is in the conduct of this part of the voyage especially, that lies the great art of the practical aeronaut, and upon which his own safety and that of his companions ultimately depends. In choosing the critical moment of the descent, and regulating his forces accordingly, much judgment and great skill are necessarily required. A certain spot, frequently at a considerable distance, is to be attained, which experience points out as best suited to the purpose, and a variety of circumstances acting separately and in conjunction must be taken into account to insure a successful issue to the attempt. The exact rate and direction of the machine at the time, and the possible variations in both, to which it may be subjected by the currents it may happen to encounter in its progress towards the earth; the amount of retardation it is sure to experience when, in the act of descending, its force of gravitation begins to operate; the quantity of gas necessary to be discharged to produce such a course as will best correspond with and satisfy these combined demands, under the restrictions of speed which a due regard to the safety and feelings of the parties necessarily imposes; all these are considerations which require to be present in the mind at once, and with such a degree of command as will enable the aeronaut in an instant to avail himself of the means within his power to provide against the consequences of any unforeseen event that may arise to derange or confound his previous calculations. The necessary acquirements for the perfect management of the descent are consequently of no ordinary nature, nor are they by any means to be met with in ordinary persons. It is not enough to entitle a man to the appellation of an accomplished aeronaut that he shall have been able to conclude his operations without breaking his neck, dislocating his limbs, or tumbling himself and his companions out of the car; to that extent all persons, with few exceptions, who have ever ascended upon their own responsibility, have shown themselves competent; and indeed, the actual peril of life or limb is so slight, that chance alone is sufficient of itself to justify the presumption of a favourable result upon that score, even in the absence of any interference whatever on the part of the manager, beyond what is necessary to determine the descent of the balloon. The mere avoidance of danger is, therefore, not the only circumstance that occupies the attention of the skilful aeronaut; a variety of other considerations, of secondary importance it is true, likewise enter into his designs. The perfect convenience and comfort of the parties, no less than their absolute security requires to be consulted : they must neither be brought to the earth with violence, jerked out of the car, dragged along the ground, hurled against buildings, nor run amongst trees; they must neither be landed in a marsh nor in a quagmire, in the middle of a wood, on the top of a house, or in the rigging of a ship, as some have had the luck to experience before now, nor decanted into the river, as has also been the fate of more than one adventurous hero, whose name figures in the annals of aerostation.

In the next place, the safety of the balloon requires and engages the solicitude of the skilful and prudent aeronaut, nor can any descent be said to have been even respectably conducted, in which the slightest injury has been allowed to accrue to that most important and valuable part of the apparatus.* This in itself involves a great many considerations. All places are by no means equally adapted for such purposes. The soil must be of such a nature as will facilitate the attachment of the balloon ; it must not be so hard that the grapnel cannot

---

* As a proof of what may be done by the exercise of proper skill, it is worth observing that the balloon which Mr. Green generally employs has already ascended two hundred and fifty-six times : one hundred and seventy with Mr. Green himself; eighty-three in charge of his son ; and three times in that of his brother, Mr. Henry Green, notwithstanding which it still remains as serviceable as ever. The great Vauxhall balloon, the unwieldy proportions of which render its management doubly arduous, has already made fifteen ascents, under the direction of the same accomplished aeronaut, among which some have been executed under circumstances of peculiar difficulty and hazard. Twice have they been deprived of the use of the grapnel by the violence of the wind, and forced to resort to adventitious expedients for the purpose of stopping the balloon ; once by the parting of the cable, and once by the actual fracture of the iron itself: it is unnecessary to observe what must have been the force of the wind by which such powerful effects were produced. Both these accidents occurred in places particularly unfavourable to the manœuvres of the aeronaut, being thickly beset with trees, and so circumstanced that had not the progress of the balloon been opportunely arrested, they would have reached the coast and been blown out to sea. With the greatest difficulty, and by the exercise of consummate skill alone, the balloon was saved from destruction. And yet with all these escapes, and the ordinary casualties of the art to boot, the silk has never so much as received the slightest puncture.

easily penetrate, nor so light that, having entered, it is unable to retain its hold; it must be free from trees or bushes, by which the silk would be sure to be lacerated, and contain a sufficiency of open, clean sward as will favour the emptying and folding of the dismembered machine as soon as its task has been performed.

Last though not least, some regard must be had for the tenants of the soil itself; much care should, therefore, be taken to avoid attempting to descend in a place where the crops are of such a nature as to suffer from the operation; a practice extremely reprehensible, not only as being the means of inflicting serious injury upon others, but likewise as tending to bring disparagement upon the art, trouble to future aeronauts, and frequently much loss and inconvenience to the parties themselves, from having their balloon seized and retained in compensation for damages, which the possession of a little skill would have enabled them to avoid.

All these are considerations which, though entirely overlooked by ordinary persons, nevertheless always enter into the calculations of the accomplished aeronaut, and require the exercise of no ordinary qualifications. Mere experience is by no means sufficient for their acquirement; for men may ascend for hundreds of times, and still keep bungling on to the end without the slightest advantage or improvement; there must be a power besides to turn it to account; a judgment to interpret its suggestions, and coolness to apply them; penetration to embrace all that is requisite at a view, and quickness in calculating the results; prudence to avoid danger, and courage to confront it;—in short, all the qualifications to a certain extent by which the skilful general is distinguished in the fields of war; and I should but ill acquit myself of my duty as an honest though humble chronicler of aerostation, were I not to mention as pre-eminent above all others in every thing which regards the practice of this delightful art, my friend, the veteran aeronaut, Mr. Charles Green. Other men there are, no doubt, in abundance, who, *under favourable circumstances*, can manage well enough to bring their operations to a close, without material injury to themselves or their companions; so far I have already said that mere chance will generally favour the attempt: it was reserved for Mr. Green to reduce these operations into a fixed and available system, and convert that chance into a matter of certainty and design.

It is not my intention to pursue the details of the descent with the same precision with which I have treated those of the earlier stages of the art. For the most part they will be found to be merely a counterpart of the preceding, differing only in the order of their occurrence, and would but weary the reader, already sufficiently so no doubt, without contributing any thing further to his stock either of information or entertainment. The few peculiarities it possesses are easily explained. Immediately upon commencing the descent, a painful impression is generally experienced in the ears, more or less acute according to the rate at which that operation happens to be conducted. I have said *generally*, because much uncertainty exists with regard to the liability to this impression; there being some in whom it is much more strongly developed than in others, while again a few there are whose physical constitution seems to exempt them from its influence altogether. The cause of the sensation is simply a renewed pressure upon the orifice of the Eustachian tube, consequent upon the passage from a rarer into a denser medium, and is so far similar to that experienced in a diving bell, although, as might be expected from the different constitution of the experiment, not nearly so strong in its indications. Indeed, in the latter, instances are not infrequent in which it has been pushed to such an extremity as to be attended with the sensation of a violent explosion in the ear, occasioned, as it is supposed, by the sudden bursting open of the valve by which the orifice of the tube is closed; producing considerable pain, nausea, and temporary, (and, in one case that I am acquainted with, even permanent) suspension of the power of hearing. This sensation continues until the descent for the time is concluded, and the equilibrium between the external air and that confined in the cavernous processes of the ear has been completely restored. Why it should not be experienced in the ascent as well as the descent of the balloon, is a circumstance most probably depending upon the valvular construction of the parts themselves, the greater facility which is afforded to the egress than the ingress of the atmospheric fluid, and con- sequently, the minor opposition encountered in the esta- blishment of the equilibrium above alluded to.

This, so far as I am aware, is the only physical impres- sion peculiar to the descent; as to the mental ones, I can only say, to speak from my own observation, that regret, intense regret, at being forced to relinquish so delightful a situation, is the only sentiment I have ever found to be an invariable attendant upon the conclusion of the aerial voyage.

But we have now no time even for the indulgence of these melancholy considerations. The balloon is already approaching the earth. The trees, hedges, roads, and other features of the rural landscape, which for some time back have been growing gradually upon the eye, have now resumed their original distinctness, and appear in quick succession, rapidly receding in our rear. Several persons now also can be distinguished, either standing in mute astonishment looking up at our approach, or hurry- ing from all directions in the hopes of being present at our descent. At length the field we have been so long aiming at, appears directly before us; the grapnel just tops the hedge, and alights immediately within it. For a few seconds it continues to drag along the ground with a succession of shocks, the violence of which the elastic cable serves considerably to abate. One, however, more forcible than the rest, at last ensues and fixes the anchor in the soil. Restricted in her progress, the balloon for the first time becomes sensible of her captivity, and seems to concentrate all the strength she possesses to effectuate her liberation. But it is all in vain. The anchor holds ; assistance multiplies in every direction ; the people run in and seize the rope ; the loss of a little more gas tames the gigantic struggler, and she stands at length secured upon the plain.

———————

In the preceding sketch, it will be perceived that I have made no account of the effects of diminished pressure upon the physical condition of the aeronaut, which some have depicted in such glowing terms, and of the exaggerated description of which I have given some examples in the commencement of the preceding Appendix.* But the truth is, that were I to speak from my own knowledge, or that of others upon whose authority I might venture to rely, and whose experience on this score is more important than my own, I should be rather inclined to dispute

* See page 102, *et seq.*

their existence altogether; at least, as obtaining at any elevation to which man, with the means he has hitherto employed, has ever been capable of ascending. In this dearth of actual testimony, all that remains for us to resort to is a circumstantial investigation of the nature of the proceeding itself, and upon these grounds it will be seen that the conclusion to which we have just arrived receives the strongest confirmation.

In the translation to the upper regions of the atmosphere, the human body, as a natural consequence of the diminished density of the medium, becomes subjected to the influence of two specific changes; namely, the remotion of pressure, and the diminished supply of oxygen gas. Now the former of these, *taken abstractedly*, I conceive to be an event of a most innoxious character, and of itself, simply, incapable of producing any effect upon the animal economy whatever. So far, indeed, is this the case, that I question if any result, seriously prejudicial to the organization of the individual, would accrue were he to be exposed to the action of a perfect vacuum in the receiver of an air-pump, providing the operation were conducted sufficiently slowly to permit the gradual escape of the included gases. This is a fact as easily demonstrated by experiments upon the inanimate as the living, and the results seem to justify our conclusion to the fullest extent. Upon the lungs, certainly, no effect whatever could be produced; the air contained therein is

always at liberty to escape, nor would any consequences
ensue from its total abstraction, so far as the simple condition of the parts themselves was concerned.

With regard to the diminished supply of oxygen, however, the case may be different; the material in question has
a specific action upon the lungs, and in certain quantities is
absolutely requisite to enable them to perform the functions
for which they are ordained. When, however, we consider how very small a portion, (not more than the five
hundred and sixtieth part) of the whole quantity contained
is consumed at each respiration,[*] and moreover, regard
the facility wherewith the organs in question adapt themselves to the changes to which, occasionally to a considerable extent, they are exposed in the ordinary course
of life, the great latitude which nature has bestowed upon
them in the exercise of functions so essential to the

[*] From the experiments of Bostock, Menzies, Sir Charles Bell,
and other physiologists, we learn that the average quantity of
air contained in the lungs of a full-grown man, is about two
hundred and eighty cubic inches, whereof forty alone, or one-
seventh of the whole, is drawn in and expelled at every ordinary respiration. Of this latter amount, according to the very
careful analyses of Mr. Davy, from one-seventieth to one-
hundreth disappears in the proceeding; assuming, however,
one-eightieth as the mean diminution produced in the quantity
actively employed, we obtain a result of half a cubic inch of
oxygen, or one-five-hundred-and-sixtieth part of the actual
contents of the lungs consumed in the process of respiration.

support of animation, we shall perceive ample grounds
for the belief that no sensible obstruction ever has or could
have been afforded to the aeronaut by the impoverishment
of the atmospheric medium at any altitude to which he
has ever been capable of ascending. These observations
are of course only intended to be applied to persons in
sound health ; it is well known to what an extent the
perceptions in this quarter become sharpened by constitutional delicacy or local disease. The circumstances under
which the ascent has been effected are likewise capable of
exercising much influence upon the physical condition of
the individual, and have, no doubt, frequently led to the
adoption of an opinion favourable to the admission of the
sensations in question as natural consequences of existence
carried on in a highly attenuated atmosphere. To this
cause, in fact, I have no doubt are to be attributed the
symptoms, slight as they are, which M. Gay Lussac describes himself as having experienced in his second excursion, when he had reached an elevation of twenty-
three thousand feet ; the greatest hitherto ever attained
by man. The only alterations which at this altitude he
was able to detect in the exercise of the functions of life,
which could in any way be imputed to the rarefaction of
the surrounding medium, was a slight increase, (amounting altogether to not more than one-third,) in the ordinary
action of the heart and lungs: considering what he says
concerning the state of his health at the time, suffering

from extreme fatigue, deprived of sleep during the whole of the preceding night, afflicted with a violent headache, and labouring, no doubt, as might be very reasonably expected, under much anxiety, not only on account of his own personal safety, but for the result of an expedition in which so mnch was at stake, and from which so much had been anticipated, the only cause of wonder is that the consequences should have been so slight as they were. Indeed I have little doubt that had it been tried, they would have been found to have been fully as great before he quitted the ground, upon his entering the car of the balloon, as at the excessive elevation whereat he was first induced to observe them.

To those who regard the difficulties experienced in the ascent of high mountains, the painful sensations and distressing symptoms to which all have more or less been subjected in the attempt to gain great elevations upon the surface of the earth, these observations and the conclusion to which they naturally conduce, may, no doubt, appear surprising. But the situations referred to are by no means analogous: in the former, a circumstance require to be taken into account which forms no part of the phenomena of aerostatic elevation; I allude to the excessive muscular action necessarily developed in the attempt; giving rise to an inordinately increased circulation, and creating an equally increased demand for oxygen gas at the very time when the natural supply, from the minor density of the atmosphere, was constantly becoming lessened. That this is the real cause of the symptoms in question, no better proof can be offered or required than the fact that all these symptoms entirely disappear the instant the exertions have been discontinued by which they were occasioned. I can only assure the reader that at an elevation in a balloon of many thousand feet above the summit of Mont Blanc, Mr. Green has assured me that not the slightest personal sensation could be detected by him different from what he would have experienced had he been sitting quietly at home in his own study.

As to the inferences which may have been drawn from the consideration of experiments upon individuals by means of an air-pump, they are not a whit more admissible as evidence of the effects of excessive atmospheric elevation than the preceding. The circumstances of the two situations are essentially dissimilar; nor would it be possible by any artificial means to render them otherwise. Either the diminution of pressure is merely local, in which case it is unnecesssary to point out the distinction; or if it be general, then does it inevitably implicate elements which do not enter into the constitution of the experiment conducted in the open air. The consumption of oxygen gas and the evolution of carbonic acid, are both essential results of the exercise of the respiratory functions, which would very soon change the nature of any experiment in closed vessels, and subject the patient to consequences from which he would otherwise be free.

Upon the whole review of the case, therefore, I have though it better to avoid all mention of the results in question, than by their admission upon dubious testimony render myself liable to the charge of having contributed to the perpetuation of error.

APPENDIX B.

The following letters have already appeared before the Public, having been inserted in the *Morning Herald* and *Times* newspapers, on the occasion of two previous ascents. I have been induced to repeat them here, merely as tending, by the descriptions they contain, to familiarize the reader somewhat more with the details of the subject in question, and furnish him with some little additional insight into the various impressions and effects peculiar to the practice of this delightful art.

NO. I. TO THE EDITOR OF THE MORNING HERALD.

SIR, October 5th, 1836.

"As you expressed a desire that I should send you some account of the proceedings of the great balloon which ascended yesterday evening from the Royal Gardens, Vauxhall, (in which I had the happiness to be a voyager,) I hasten to comply with your request, although, from the frequency of such exhibitions, and the numerous statements on the subject which have appeared of late years, little can be expected, either interesting or new, to be added to the general stock of information already before the public.

The inflation of so large a balloon, and the general preparations for the ascent upon so unusual a scale,* were, no doubt, the grand features of the exhibition upon the present occasion. In order to make certain of being ready to start at the hour appointed, these preparations had been commenced as early as six o'clock in the morning; and by the time the public had arrived, upwards of fifty thousand cubic feet of gas, or nearly two-thirds of the whole contents of the balloon had been already

* Historical accuracy obliges me to observe that the magnitude of the Vauxhall balloon is by no means its distinguishing characteristic, or that it can lay any claim to pre-eminence on that score, when placed in comparison with some of those by which it has been preceded. Indeed, in point of size, it much falls short of any of those by which the first dawn of the art was signalised, and in which the ascensive agent was atmospheric air, rarefied by the application of artificial heat. The Vauxhall balloon is a spheroid of about sixty feet in height, and capable of containing about eighty thousand feet of gas. The first balloon which ever was made, with a view to the support of weight, was likewise a spheroid of seventy-two feet in altitude, and, in a preliminary experiment, was found adequate to support the weight of eight men. This balloon was immediately destroyed by an accident, and almost as quickly replaced by another of the same form and dimensions. The third balloon of which I have any note, was that in which M. Pilatre de Rosier and the Marquis d'Arlandes made the first aerial voyage, and was seventy-four feet in height, and forty-eight in its extreme lateral extension. The fourth of which I am aware was even still more stupendous; it was no less than one hundred and thirty feet in height, and one hundred and five in breadth; and contained the enormous volume of 540,000 cubic feet of gaseous contents, (nearly eight times the amount of the Vauxhall balloon.) It took fifty men to retain it, and ascended at Lyons, on the 19th of January, 1784, under the direction of its maker, the celebrated Joseph Montgolfier, with a charge equal to about two tons and a half, independent of its own weight and that of its apparatus. In Italy, the first balloon that ever ascended was a sphere of sixty-eight feet in diameter, and was launched at Milan, the 25th of February, 1784, with three persons, and the proper quantity of ballast and other equipments. It is in the construction of the Vauxhall balloon, its beautiful form and exquisite workmanship, that lies its real claim to notice; and in which particulars it certainly has hitherto, and most probably for a long time to come will remain unequalled.

thrown in. Were it not for the sake of giving the fullest effect
to the size and appearance of the balloon, this quantity alone
would have been amply or perhaps more than sufficient for the
purposes of the ascent. As the gas naturally expands in proportion as it ascends into a rarer medium, whatever exceeds
the quantity which it would take to fill the balloon, under the
diminished pressure of the greatest altitude to which it is required to proceed, must evidently be lost in the progress of the
ascent, and might as well have been altogether dispensed with
from the beginning, so far as its utility in aiding the buoyancy
of the balloon is concerned. With a view, however, to displaying the form of the gigantic sphere in all its majesty, and affording a spectacle worthy of its real dimensions, the inflation was
continued until the whole was filled, and upwards of seventy-
five thousand cubic feet of carburetted hydrogen gas had been
transferred from the reservoirs of the neighbouring company,
(the London gas-works,) into the silken vessel destined for its
reception. And a splendid spectacle it certainly did afford
when thus provided ; extending upwards of sixty feet in height,
far out-topping every thing in its neighbourhood, and almost
appearing to fill the whole space devoted to the operations of
the ascent. Upon a par with this stupendous sphere were all
the other accessories to the ascent ; the car, hoop, netting,
cable, and what must have been still more gratifying to the proprietors, the company who had assembled to behold it, and by
whom the hour of the ascent was now impatiently expected. At
length the process of inflation being completed, and all things
duly prepared, the balloon, with nine persons in the car, quitted
the earth, at about a quarter after four o'clock, and, as you may
have observed, immediately assumed a course which shortly
brought it above the eastern portion of the metropolis ; across
which it rapidly passed in such a direction, and at such an
elevation, as afforded to the spectators, both above and below,
the most complete and favourable display of the objects at the
time most interesting to each.

" It would be impossible even cursorily to note the numerous points of view which captivated and almost *confused* the attention in the transit across this magnificent and constantly changing prospect. The appearance of a vast and voluminous surface unrolling itself to the eye; a city contracted to a span, yet distinctly exhibiting every feature at one glance, and with all the adjuncts of animation, which no picture can convey, and scarcely words adequately explain; the Thames throughout its long course of windings, for upwards of fifty miles on either side the centre of our lofty line of vision, covered with its myriads of appropriate tenants, ships, barges, and even wherries, microscopically portrayed; the river Lea, contributing its mite through a series of tortuous deviations, now shrivelled to a mere silver thread, with its various reservoirs, glittering like fragments of a broken mirror scattered upon the green sward; the docks, canals, bridges, buildings, streets, roads, parks, and plantations, appearing to recede in peristrephic order, to make way for new prospects, and give occasion to renewed admiration and applause; all these, and a thousand others nameless, though equally deserving of being named, combined to afford a prospect which to be appreciated must be seen, and once seen can never fail to be remembered.

" Notwithstanding the elevation at which the balloon passed over the city, the distant hum of voices, the subdued murmur of active population, and occasionally the reiterated clamour of the multitude cheering the machine along its airy course, was distinctly audible amid the very remarkable stillness which qualified the surrounding medium.

" Were I to institute a comparison with any of the more familiar methods of delineating nature, for the purpose of giving to those who never witnessed it, an idea of the appearance which the scene below us, especially the city and its immediate vicinity, bore at this period of our voyage, I should be inclined to say that it more nearly resembled in its effect the mimic representations of some vast camera-obscura, in which,

with all the fidelity of nature itself, the most rigorous observance of proportion between the size and motion of the various
bodies, is combined with the most perfect delineation of their
minutest forms throughout every scale of decreasing magnitude, until they no longer continue to be discernible.

" After passing over the metropolis, the balloon continued its
course in the same direction in which it first started, rapidly
traversing the north-eastern division of the confines of London,
until it had reached that portion of Epping which is known as
Hainault Forest, displaying a varied landscape of towns, woods,
water, and cultivated inclosures, only to be matched under
similar circumstances, and by the same mode of conveyance
and exhibition. At this point in its progress, the course of the
balloon slightly deviated from its original direction about a
point and a half to the eastward, maintaining a nearly uniform
elevation of about 2,500 feet, owing to which circumstance, we
not only obtained a clearer view of the world beneath us, (which
from the dense clouds that completely canopied the sky during
the whole day, we should have altogether escaped, had we proceeded to ascend higher,) but also no doubt exhibited to the
numerous spectators over whom we passed, and whom we could
plainly perceive stopping to examine us, a more interesting
subject of observation, than under other circumstances we
should have been calculated to afford. Indeed, I have no
doubt that, to many of those below, too minute to attract our
notice, we must have occasioned much interest and some fear, if
we might judge from the conduct of the numerous flocks of ducks
and geese, the denizens of the various marshes and pools over
which we passed, which might be seen running in every direction, as we approached, to seek a shelter from the enemy, that
appeared like a stupendous falcon hovering above them. The
very cows in the neighbouring fields appeared to view our proceedings with considerable distrust, and though, no doubt, with
less defined notions of danger, seemed to acknowledge the pre-

sence of an enemy, and exhibited as much confusion as it was possible to express by tails erect and accelerated paces.

It is astonishing with what a degree of apprehension and alarm, all animals, birds especially, appear to regard the presence of the balloon, and what singular shifts, each, according to their several natures and means of refuge, adopt for the purpose of avoiding its approach. I have frequently observed with much interest the proceedings of a colony of rooks, acting under the influence of these alarms, at a time when the progress of the balloon happened to be more than ordinarily accelerated by a stiff breeze. Aroused from their airy habitations, and unable, no doubt, at first, to fathom the powers and intentions of this novel apparition, they seem to content themselves with advancing leisurely, by short stages, occasionally stopping whenever they have reached a certain distance, and again renewing their retreat as the balloon begins to diminish the interval which they seem desirous to maintain between them. Finding, however, the inutility of these desultory efforts, and perceiving the mighty object of their terrors gradually gaining ground upon them, they soon abandon their former line of conduct, and taking the direction of the balloon, seek in a continuous flight to evade the antagonist with which they have apparently no other mode of contending. In this manner they persist, vainly increasing in their exertions, until at length the balloon, having attained a position directly above them, they suddenly stop short, seem to hesitate for a moment in dismay, and finally, after a brief but stormy debate, adjourn in confusion, scattering themselves in every direction over the surface of the earth beneath.

"Upon arriving over the neighbourhood of Chelmsford, it was thought proper, owing to the approach of dusk, to prepare for descending, in order that we might be able to avail ourselves of the remaining daylight to effect the package of the balloon. Accordingly Mr. Green, having previously prepared and suspended the grapnel, drew the valve-string, and reluctantly we commenced quitting our airy situation. The descent of the balloon was now made apparent to our senses by the slight pressure on the ears, occasioned by our rapid entrance into a denser medium, a sensation exactly similar to that experienced in the diving bell, when the machine is immersed to any depth beneath the surface of the water. From this effect, however, the attention was speedily withdrawn by the grapnel coming in contact with the soil, and giving us the first signal of our return to earth. The force of the wind, of which, while suspended in the air, we had been totally unconscious, now, for the first time, began to make itself felt. The grapnel, unable for want of a sufficient hold, to arrest the rapid progress of the balloon, continued to trail on the ground for a short space, with a succession of shocks, which would have been severely felt but for the adoption of the elastic India-rubber cable, a contrivance which fully answered the purposes for which it was designed, and cosiderably mitigated the unpleasant consequences hitherto attendant upon this stage of the descent, especially when accompanied by a moderate gale of wind, such as was experienced in the present instance. After, however, rapidly traversing one or two inclosures, and passing, with some resistance, through a hedge, the grapnel became firmly fixed, and the balloon, speedily disburdened of a further portion of its gas, finally reached the earth, in a field between Chelmsford and the village of Writtle, in about fifty-five minutes after its departure from Vauxhall Gardens, having in that time traversed a space which, in a direct line would have measured about thirty-two miles, but which, allowing for the arch formed by the ascent and descent, and the further deviation of its course from a right line, would have given a result more nearly approaching to forty miles. Scarcely had the balloon touched the earth when we were surrounded and literally *involved* in the numbers who continued to flock in from all quarters, on foot, in carriages, and on horseback, all anxious to lend their assistance to secure the balloon, with a zeal which their very numbers and eagerness almost rendered abortive. By the timely interference, however, of Mr. Bramston, M.P., Mr. Bacon, Mr. Guy, the chief constable of the town, and others, some order was produced, and the balloon and other accessories safely packed in the car, and forwarded by a van to London.

"As we were proceeding to Chelmsford in chariots which had been despatched for the purpose, the whole neighbourhood, with one accord, appeared to have turned out to greet us; and preceded by a band of music, which, to our surprise, had been summoned on the occasion, and surrounded by a concourse which certainly did not fall much short of four thousand persons, of all sexes ranks, and ages, we entered the town, whence, after some suitable refreshment, and a proper acknowledgment of the honour conferred upon us, we started for London, where we arrived at twelve o'clock this morning.

"It was rather a singular coincidence that at the different places where we stopped to change horses on the road, we found that we had been preceded only a few minutes by Mrs. Graham, who, though under more fortunate circumstances, was likewise returning home for the first time after her late accident.

"It would be most unjust were I to conclude without expressing in my own name and that of my companions our full sense of Mr. Green's deportment during the whole proceeding, and adding our own to the testimonies of his unrivalled abilities as an aeronaut, and his excellent qualities as a companion, which he has never failed to exact from all persons who have had the good fortune to attend him in his excursions.

"M. M."

NO. II.   TO THE EDITOR OF THE TIMES.

SIR,   October 18th, 1836.

" Perhaps I ought to apologize to you and your readers for
so soon troubling them upon the same subject, and one which,
from the frequency wherewith it has been treated of late, may
very reasonably be considered as almost exhausted.   The great
variety, however, of the aspects under which nature exhibits
herself in such situations, and the novelty of the manner in
which even her most ordinary features are displayed to those
who avail themselves of such a mode of examining them, will at
all times leave room, even for the most superficial observer to make
some remarks which have escaped the comments of former
aeronauts, and to note some peculiarities which distinguish each
successive ascent from all those which have preceded it.   Scientific experiments are, of course, out of the question in an ascent
which has not been conducted with an especial view to such
ends, and where the elevation attained, (which is, in fact, the
chief grounds for its employment in such purposes,) was not
calculated to admit of any beyond those of the most usual and
common-place description.   So manifold, however, are the operations of nature, and so replete with interest even the most
insignificant of her works, that no two ascents can ever be
said to be so perfectly alike that something may not remain to
be told to interest the general reader, and excuse the recurrence
of a subject which must, yet for a long period, continue to be
classed among the most striking novelties of an enterprising
age.

" At twenty-five minutes to four the balloon and car, containing nine persons, rose majestically from the ground, and,
assuming at the first a south-westerly direction, rapidly traversed
the extremity of the firework gallery, immediately and closely
sweeping over the heads of the persons who had collected there

Q

for the purpose of witnessing the ascent. As soon, however, as she had reached a slight elevation, her ascensive power quickly prevailed, and in a few seconds she was involved in the clouds which impended at a slight distance above the surface of the earth.

" Although the day might be considered as generally unfavourable to aeronautical display, yet was it not without its advantages, especially to those whose previous experience in such scenes had been confined to a clear atmosphere or an unclouded sky. The vast extent of vapour which canopied the earth, and ultimately excluded that object from our view, if in one point it was calculated to detract from the beauty of our prospect, by depriving us of one great and usual source of admiration, in another contributed to the interest and majesty of the scene by the novel aspect under which it presented the altered face of nature to our senses.

" Scarcely had we quitted the earth before the clouds, which had previously overhung us, began to envelop us on all sides and gradually to exclude the fading prospect from our sight. It is scarcely possible to convey an adequate idea of the effect produced by this apparently trivial occurrence. Unconscious of our own motion from any direct impression upon our own feelings, the whole world appeared to be in the act of receding from us into the dim vista of infinite space; while the vapoury curtain, like similar phenomena represented on the stage, seemed to congregate on all sides and cover the retreating masses from our view. The trees and buildings, the spectators, and their crowded equipages, and finally, the earth itself, at first distinctly seen, gradually became obscured by the thickening mist, and growing whiter in their forms, and fainter in their outlines, soon faded away " like the baseless fabric of a vision," leaving us, to all appearance, stationary in the cloud that still continued to involve us in its watery folds. To heighten the interest and maintain the illusion of the scene, the shouts and voices of the multitude whom we had left behind us, cheering the ascent, continued to assail us, (long after the interposing clouds had effectually concealed them from our eyes,) in accents which every moment became fainter and fainter, till they were finally lost in the increasing distance.

"Through this dense body of vapour, which may be said to have commenced at an altitude of about 1,000 feet, we were borne upwards to perhaps an equal distance, when the increasing light warned us of our approach to its superior limits, and shortly after the sun and we rising together, a scene of splendour and magnificence suddenly burst upon our view, which it would be vain to expect to render intelligible by any mode of description within our power : pursuing the illusion which the previous events had been so strongly calculated to create, the impression upon our senses was that of entering upon a new world to which we had hitherto been strangers, and in which not a vestige could be perceived to remind us of that we had left, except the last faint echo of the voices which still dimly reached us, as if out of some interminable abyss into which they were fast retreating.

"Above us not a single cloud appeared to disfigure the clear blue sky, in which the sun on one side, and the moon in her first quarter upon the other, reigned in undisturbed tranquillity. Beneath us in every direction, as far as the eye could trace, and doubtless much farther, the whole plane of vision was one extended ocean of foam, broken into a thousand fantastic forms ; here swelling into mountains, then sinking into lengthened fosses, or exhibiting the appearance of vast whirlpools ; with such a perfect mimicry of the real forms of nature, that, were it not for a previous acquaintance with the general character of the country below us, we should frequently have been tempted to assert without hesitation the existence of mountainous islands penetrating through the clouds, and stretching in protracted ranges along the distant verge of our horizon.

"In the centre of this hemisphere, and at an elevation of about 3,000 feet above the surface of the clouds, we continued to float in solitary magnificence; attended only at first by our counterpart—a vast image of the balloon itself with all its paraphernalia distinctly thrown by the sun upon the opposite masses of vapour, until we had risen so high that even *that*, outreaching the material basis of its support, at length deserted us; nor did we again perceive it until, preparatory to our final descent, we had sunk to a proper elevation to admit of its reappearance.

"Not the least striking feature of ours and similar situations is the total absence of all perceptible motion, as well as of the sound which in ordinary cases is ever found to accompany it. Silence and tranquillity appear to hold equal and undisputed sway throughout these airy regions. No matter what may be the convulsions to which the atmosphere is subjected, nor how violent its effects in sound and motion upon the agitated surface of the earth, not the slightest sensation of either can be detected by the individual who is floating in its currents. The most violent storm, the most outrageous hurricane, pass equally unheeded and unfelt; and it is only by observing the retreating forms of the stable world beneath that any certain indication can be obtained as to the amount or violence of the motion to which the individual is actually subjected. This, however, was a resource of which we were unable to avail ourselves, totally excluded as we were from all view of the earth, or any fixed point connected with it.

Once and only once, for a few moments preparatory to our final descent, did we obtain a transitory glimpse of the world beneath us. Upon approaching the upper surface of the vapoury strata, which we have described as extending in every direction around, a partial opening in the clouds discovered to us for an instant a portion of the earth, appearing as if dimly seen through a vast, pictorial tube, rapidly receding behind us, variegated with furrows and intersected with roads running in all directions; the whole reduced to a scale of almost graphic minuteness, and from the fleecy vapour that still partially obscured it,

impressing the beholder with the idea of a vision of enchantment,
which some kindly genius had, for an instant, consented to disclose. Scarcely had we time to snatch a hasty glance, ere we
had passed over the spot, and the clouds uniting, gradually
concealed it from our view.

" After continuing for a short space further, in the vain hope
of being again favoured with a similar prospect, the approach
of night made it desirable that we should prepare for our return
to earth, which we proceeded to accomplish accordingly.

" It is in the management of the descent under circumstances
similar to those which characterized the present occasion that
the utmost skill of the aeronaut is principally displayed. The
low position of the clouds, resting almost upon the earth itself,
precludes the possibility of observing the nature of the ground
until it would, without the exercise of the greatest judgment,
be impossible to avoid completing the descent, however unfavourable the country might eventually prove for such a purpose.
To all this detail, however, Mr. Green proved himself perfectly
competent ; the balloon gradually descended into the cloudy
region below us, and became involved for a minute or two in obscurity ere we perceived ourselves slowly emerging over a large
tract of ploughed land particularly well adapted to our design.
Scarcely had another minute elapsed before the grapnel reached
the ground, on which it continued to drag with some resistance
for a short space, until it took a firmer hold of the soil ; when two
gentlemen (one of them Mr. Cumberlege, the clergyman of the
neighbouring district), who were riding with some ladies, perceiving our situation, leaped from their horses and with a zeal which
merited our thanks lent their aid to secure the grapnel more firmly.
More persons shortly after arriving, the balloon was finally
brought to the earth, and we effected our landing in a common
called Billington-fields, in the parish of Leighton Buzzard, about
two miles beyond that town and about forty-eight from the
gardens at Vauxhall ; having employed about an hour and three

quarters in the voyage, upon a nearly uniform course of north-west by north, and at a nearly uniform elevation of about 5,000 feet above the level of the sea.

M. M.

NO. III. TO THE EDITOR OF THE MORNING HERALD.*

SIR,   July 23rd, 1837.

IN consideration of the forthcoming experiment, which I see announced for to-morrow, perhaps the following observations upon the subject of the parachute in general, and the merits of the two different systems which are now about to be practically illustrated, in particular, may not be thought unworthy of notice.

The principal of the parachute is so extremely simple, that the idea must no doubt have occurred to many persons, of whom history, however, has failed to preserve a record. Even in the distant and half-civilized regions of Siam, Father Loubere in his curious account of that country, published nearly two

* This letter was addressed to the Editor of the Morning Herald in contemplation of an experiment announced for the following day, namely, the descent of Mr. Cocking in a parachute, stated to be on a new and improved principle of construction. In consequence of the lateness of the hour at which it was despatched to the office, (two o'clock in the morning of publication), it was found impossible to do more than merely to state the results. In its republication, I have given the whole as it originally stood, without the slightest alteration or amendment, except the addition, in notes, of certain calculations and observations which the nature of the public journals rendered inadmissible, but which, to those who happen to take an interest in these matters, may not prove unacceptable.

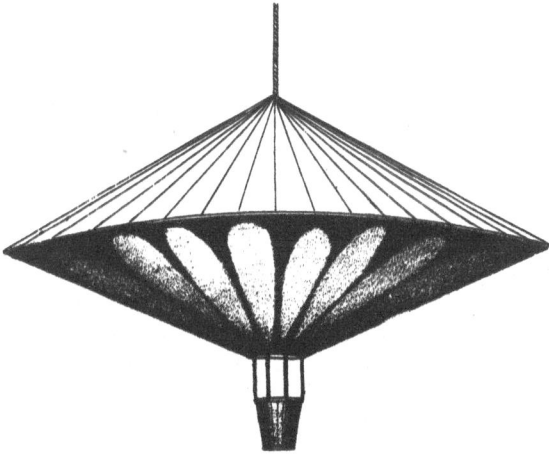

PARACHUTES OF M. GARNERIN AND M.ʳ COCKING.

centuries since, makes mention of one who was wont exceed-
ingly to divert the court by his exploits in descending from
great heights through the aid of such an instrument; a practice
which we have reason to believe was not confined to that alone
of all the countries of the East. In Europe, however, no
notice appears of any attempt to employ the parachute as a
preventive against a too rapid descent through the atmosphere,
till the year 1783, when a gentleman of the name of le Normand
first practically demonstrated its efficiency, by letting himself
down from the windows of a high house at Lyons, of which city
he was a native. The views of M. le Normand with regard to
its employment were, however, very limited, and do not appear
to have extended further than its adoption as a means of escape
from fire; nor was it till some time after, that the ingenious
and speculative Blanchard first conceived the idea of applying
it as an adjunct to the then new and interesting art of aerostation. This design he endeavoured to put into execution in
an ascent which he executed at Basle, in the year 1793; having
previously satisfied himself of its security by letting down dogs
and other animals from various heights in the course of several
aerial excursions undertaken from Strasburg, Lisle, and other
places during the several preceding years. In attempting to
repeat the experiment upon himself, however, he was less fortunate; owing to some mismanagement, his machinery failed in
its effect, and, coming to the ground with too great rapidity, his
leg was broken in the fall.

To André-Jacques Garnerin who next followed in the career
of the parachute, is due the merit, such as it is, of having been
the first who ever succesfully descended from a balloon by the
aid of that machine. This he accomplished in an ascent from
Paris, on the 21st of October, 1797, in the presence of the
court of France, and of an immense concourse of people who
had assembled to witness the adventurous experiment. At the
height of about 2,000 feet the act of separation was effected,
and the balloon and parachute immediately started off in

opposite directions. The former, however, was soon lost sight of, all eyes being involuntarily directed towards the descending mass, and all interest centred in the individual it contained. For a few seconds, the consummation of his fate seemed to be altogether inevitable, the parachute obstinately retaining the collapsed position in which it had originally ascended. All of a sudden, however, it burst into its proper shape, and the downward progress of the adventurer appeared at once to have been arrested. The fears of the spectators now began to assume another aspect; the moment the parachute had expanded, the car of the aeronaut, which was suspended about twenty feet below it, shot out on one side with an impetus that almost brought it upon a level with the rest of the apparatus, and for an instant seemed to threaten the subversion of the whole. Recovering itself, however by its force of gravitation, it soon re-descended, and swinging round to the opposite quarter, commenced a series of violent oscillations, which for a considerable time seemed to render the issue of the experiment a matter of much uncertainty. As he approached the earth, however, these gradually became fainter, and although they never entirely disappeared, soon ceased to excite the immediate apprehension of his friends. At length, in about twelve minutes, he reached the ground, and was released from the parachute, without having experienced other injury than a feeble shock at the instant of collision, and a slight nausea which shortly after supervened, occasioned it is supposed by the unsteady nature of the movement to which he was subjected in the descent.

Shortly after this, Garnerin proceeded to England, where he made his third essay, in an ascent from North Audley Street, on the 21st of September, 1802, being the only one of the kind hitherto ever exhibited in this country. Since that period the parachute has frequently been made use of, both by himself and others in various parts of the continent, always, however, for the purposes of public exhibition; nor indeed, am I aware of any instance, except one, in which any absolute advantage has ever accrued from its employment: I allude to the case of Jordaki Kuparento, a Polish aeronaut, who, on the 24th of July, 1804, ascended from Warsaw, in a *montgolfière*, or fire-balloon; when at a considerable altitude in the sky his balloon became ignited; being provided, however, with a parachute, he was enabled to descend in safety.

The principle upon which all these parachutes were constructed is the same, and consists simply of a flattened dome of silk or linen, from twenty-four to twenty-eight feet in diameter. From the outer margin all around, at stated intervals, proceed a large number of cords, in length about the diameter of the dome itself, which being collected together in one point, and made fast to another of superior dimensions, attached to the apex of the machine, serve to maintain it in its form when expanded in the progress of the descent. To this centre cord likewise, at a distance below the point of junction, varying according to the fancy of the aeronaut, is fixed the car or basket in which he is seated, and the whole suspended to the net-work of the balloon in such a manner as to be capable of being detached in an instant at the will of the individual, by cutting the rope with a knife, or, still better, by pulling a string communicating with a sort of trigger or pivot by which it is made fast above.

In the choice of the form of the parachute, its original inventors were chiefly guided by the desire to obtain the greatest atmospheric resistance consistent with a given extent of surface; and although the form they did adopt may not be that which answers *exactly* to this description, yet it falls short of it so little as to more than compensate the deficiency by the other advantages which it affords.

Two objectionable circumstances, however, are generally found to attend the employment of the parachute as here described; namely, the length of time which is wont to elapse before it becomes sufficiently expanded to arrest the fall of the individual, and the violent oscillatory movement which almost invariably accompanies the descent.

In order to obviate these deficiencies a variety of plans were proposed at different times, amongst which is that now shortly about to be tried, and for which I perceive the proprietors of Vauxhall Gardens claim the merit of originality. The idea, however, is a very old, and a very common one, although from certain inherent deficiencies, the practical cultivators of the art have declined adopting it. It was published in Paris nearly forty years ago; revived in England by Sir George Cayley, and communicated by him, with other interesting notices upon aerostation, if I mistake not, to the twenty-fourth volume of Nicholson's Journal; it was subsequently more fully developed and improved by Mr. Kerr, by whom it was in several experiments practically and publicly illustrated, and is finally detailed in the Encyclopædia Edinensis, at the conclusion of the article headed " Aeronautics."

The principle of the plan alluded to, is simply an inversion of the preceding ones, in which the surface of least resistance is made to descend foremost, and so contrived, as at all times to remain in a state of expansion. The precise form of that which is now announced for experiment is an inverted cone, somewhat flattened, to the apex of which is attached the car of the adventurer. The chief objects of this arrangement I have already stated to be the correction of the oscillatory motion, and the insurance of the speedy action of the machine after its detachment; to the former of these, its shape was intended to conduce : to the latter its state of permanent expansion. And yet in seeking to obviate the irregularities in question by any modification in the form of the parachute, a great error has been committed, which nothing but an ignorance of their real cause could ever have occasioned. Indeed, these oscillations seem very much to have puzzled the aeronautical world, both here and elsewhere, and yet the grounds upon which they are accountable are extremely simple. Entirely independent of the form, the aberrations in question are merely the consequence of a first irregularity impressed upon the machine by the unequal expansion of its parts. In the act of opening, it is next to impossible that all the gores of the capacious dome should in the same moment attain the same degree of elevation; the side which is first opened to its full extent, receives the first impression of resistance; the machine is thrown out of its equipoise; the irregularity which it first assumes becomes quickly transferred to the other side by the gravitation of the appended weight, and a reciprocal interchange of forces thus becomes established, which the atmosphere possesses but too little consistence speedily to subdue. Any attempt to correct these derangements by a modification of the *form* of the parachute is extremely futile; but to endeavour to do so in the way proposed is worse than futile: it is really to sacrifice the very principles of the machine to the attainment of an end to which the condition in question does in no way conduce. By a course of calculation founded upon the admitted axioms of dynamics (all of which are in fact the results of actual experiment), we learn that the resistance upon the base of a cone (supposing it a plane surface), is to that upon its oblique presentation in the proportion of unity to the sine-squared of half the vertical angle.*

---

* The following investigation of the comparative resistance of fluids to bodies of different forms, is by my friend W. C. Ottley, Esq., Fellow of Caius College, Cambridge, and will not be examined without interest, especially, considering the disastrous event, the probable occurrence of which it was originally undertaken to elucidate.

"It is usual to calculate the resistance of fluids on bodies in motion upon the hypothesis of the particles of fluid leaving the surface of the body without impediment immediately after impact. This hypothesis is evidently incorrect in practice, inasmuch as the particles of air reflected from the surface must more or less interfere with those in progress towards it. It will be easily conceived, that this effect must be the greatest when the surface on which the air impinges is concave, and that consequently it must generate a kind of compression in the concavity, which must much increase the whole effect of the resisting fluid. The increase of resistance arising from this cause will diminish gradually as the concavity diminishes, and will still be considerable when the surface is a plane; but whenever it becomes convex the effect arising from this

Supposing the apex of the cone to be an angle of 120 degrees, (from which I have heard it is not far removed), this proportion, it will be seen by reference to the note below, would stand in

source becomes inappreciably small, from the facility with which the particles glide off after impact. These observations apply to that part of the effect which is disregarded in the mathematical calculations of he resistance of fluids; but it will be presently shown, that even apart from these considerations, the resistance of the air upon a convex surface is considerably less than that upon a plane. To the mathematical reader this will be at once apparent; for the instruction of others, however, it may be as well to observe, that in the case of the plane the impact is direct, and consequently the whole momentum of the particles of air is exerted in resisting the advance of the body, whereas in the case of the convex surface of a cone, the impact being oblique, only a certain portion of that force becomes effective in opposing its progress through the atmosphere.

It is true, that the effect of the friction of the air against the convex surface of the cone would in some measure tend to increase this resistance, and that this effect in a cone with a very acute angle might considerably modify the required calculation; but in the case of a cone whose vertical angle is obtuse this effect may safely be disregarded, as more than counteracted by the circumstances just alluded to.

Proceeding to calculate the difference between the resistance on the convex surface of a cone and on its base, we shall find it considerable, and if, besides, we take into account the effect of the interference of the reflected with the impinging particles of air, we shall see how much there is in favour of the concave parachute.

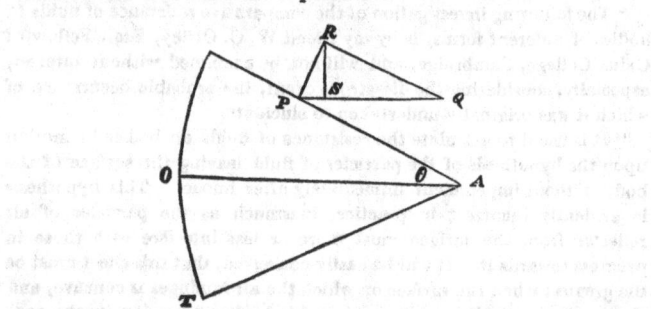

Let PQ represent the force of a particle of air impinging upon the surface of the cone APT, in the direction of its motion. Then draw PR at right angles to the surface of the cone, QR at right angles to PR, and RS perpendicular to PQ.

numbers exactly as 4 is to 3 ; thus indicating a loss of one quarter of the effect which would have been afforded by an horizontal area equal in extent to its base. Assuming, therefore, the superficial contents of this latter to be 908 square feet, (as would be the case were its radius 17 feet, than which however it is something less), the power of the parachute in question to retard the fall of the individual, would only be equal to that of a circular plane whose surface was 681 square feet.

The terminal velocity of such a parachute, or the rate at which it would reach the ground, is easily computed. From the experiments of Ferguson, Rouse, Smeaton, and others, on the accuracy of which the greatest reliance may be placed, we learn that the force exerted by the atmosphere in motion at the rate of one mile per hour, against a plane at right angles to the direction of its course, is in the ratio of . 005 of a pound avoir-

Then the effective resolved part of the force QP on the surface of the cone $=$ PR $=$ PQ sine RQP $=$ PQ sine PAO $=$ PQ sine θ; and the part of this force resolved in the direction of the motion of the cone, and therefore effective in resisting its advance, will be represented by PS $=$ PR sine PRS $=$ PQ sine $^2\theta$, since PR $=$ PQ sine θ.

Now, observing that the compactness of the impinging particles of fluid on any given portion of a surface will be proportional to the sine of the angle of inclination, (sine θ), and putting ϱ to represent the direct resistance of the air on a unit of surface, and dS to represent an elementary portion of the surface of the cone, we have $dR = \varrho$ sine $^2\theta$ dS; and if $y = ax$ be the equation to the line AP, and s the length of the line, we have $dS = 2 \pi y \, ds = 2 \pi y$ sec. $\theta \, dx$

$$\therefore \; dR = \varrho \text{ sine } ^2\theta \; 2 \pi y \text{ sec } \theta \; dx = 2 \pi \varrho \text{ sine } ^2\theta \; ydy$$
$$\therefore \text{ integrating } R = \pi \varrho \text{ sine } ^2\theta \; y^2.$$

Now the resistance on the base of the cone $= \varrho$ multiplied into the surface of the base, because the impact is direct; $\therefore$ R $= \varrho \pi y^2$. *Hence the resistance on the surface of the cone is to the resistance on the base as sine ^{2} ½ angle of the cone is to unity.*

Thus, if the vertical angle of the cone were 90°, then ½ angle $=$ 45°; now sine 45° $= \sqrt{\frac{1}{2}} \therefore$ sine ^{2} 45° $=$ ½ $\therefore$ the resistance on the base of such a cone is double that on the surface. Again, supposing the vertical angle of the cone to be 120°, then ½ angle $=$ 60°; now sine 60° $= \sqrt{\frac{3}{4}} \therefore$ sine ^{2} 60° $=$ ¾ $\therefore$ the resistance on the base would in this case be to that upon the surface as 4 : 3."

dupois for each square foot of surface; which force we are further aware increases directly as the squares of the velocities under which it is exercised. It is almost unnecessary to observe, that whether the atmosphere impinge upon the surface or the surface upon the atmosphere, the effect, as far as the question of resistance is concerned, is precisely the same.

Now, the weight of the above apparatus, inclusive of the individual himself, cannot, I am convinced, be safely computed at less than 500 pounds. I am aware, that in the public announcements the weight of the parachute is stated to be but 223 pounds;* from the little acquaintance, however, which I have had with such experiments, I am perfectly satisfied that no machine of the alleged dimensions could be constructed, with the slightest regard to stability, in which the weight was under 350 pounds; and if to this we add 150 pounds for the individual himself, his ballast, and other equipments, I have no doubt we shall rather fall short of than exceed the reality. Upon this supposition, therefore, and assuming, as above calculated, a plane of 681 square feet to be equivalent to the parachute in question, we shall find a force of . 703 of a pound exerted upon every square foot; indicating, according to the scale before laid down, a rate of motion of about 12 miles an hour, or nearly 18 feet in a second.†

* Some idea may be had of the loose manner in which a transaction, involving no less than the life of a man, was conducted, when we observe that in the public announcement, from which alone the world could derive any information on the subject, the weight of the entire apparatus, including that of the individual himself, was stated to be but 393 pounds; whereas from the evidence taken before the coroner, in the inquest upon the body of the unfortunate victim, as will be seen further on, it was made apparent that, after deducting 170 pounds, (which it seems was the weight of Mr. Cocking), there still remained 413 pounds to be laid to the account of the apparatus alone; very nearly twice as much as that at which it was originally computed.

† The calculations according to the formula of Dr. Hutton, which here followed in the original, give a result so nearly coinciding with that

To those who are not in the habit of forming an estimate of consequences upon data of the above nature, it will serve to give some notion of the force developed in such a proceeding, merely to suggest the consideration of the shock they would receive were they to be launched unprotectedly against a solid wall from the top of a vehicle travelling continuously at the rate of 12 miles an hour.

Nor is this a result peculiar to the above alone of all parachutes upon the same construction, or one which any arrangement of its condition with regard to weight and size, could ever enable it to avoid. Owing to the perverse nature of the principle upon which it is contrived (all the forces which it encounters in its employment, acting in direct opposition to the maintenance of its proper form), a degree of strength becomes necessary in its construction totally incompatible with the requisition of weight, essential to the proper regulation of its descent. This is an inconsistency which it is impossible to reconcile by any means within our power. There is a certain limit in nature to the strength of materials compared with their weight, which all the art of man can neither alter nor extend: in some cases this limit is very speedily attained; and I think it would not be difficult to prove that in this particular instance it falls far short of what would be necessary to answer the purposes in view.

With such an obstacle to contend with, I have no hesitation in declaring that no parachute can ever be constructed upon the principle in question that shall be capable of retarding the fall of man within the restrictions of speed necessary for his final preservation. No argument in contravention of this position, drawn from a consideration of experiments upon a smaller scale, is at all admissible. In comparative experiments of this

above stated, that we have thought it unnecessary to repeat them here. The terminal velocity, as computed upon these grounds, would have been $19\frac{1}{15}$ feet in a second; somewhat greater than that deduced by the above.

nature there are certain elements which cannot be made to keep pace with the rest, and which remaining always the same, utterly invalidate any analogy which it might be thought proper to institute between them. So long as the service required of them falls within a certain limit, there is no doubt of the success of their employment; the moment it passes that limit, one or other of these fixed principles begins to give way; nor can its place be either dispensed with or supplied by any modification of the rest: on the contrary, any attempt to resort to such a remedy only tends to multiply the forces by which that fixed principle itself is really subdued.

If this is true in cases where the modifications alluded to are not necessarily more than are required for the end to which they are sought to be applied, as for instance, where an increase in the quantity of material is merely made to supply a deficiency in its strength it is doubly true where their introduction absolutely gives rise to circumstances by which a further increase in their amount is imperatively required. The manner in which this operates in the present case, will appear the more readily when we consider that all the modifications in question, involving the increase of weight for the purposes of strength, are referrible to the great hoop or upper frame-work of the machine, tending directly to the derangement of its equipoise, and calling for the further addition of weight in another quarter, where it not only conduces nothing towards strength, (the want of which it was originally introduced to supply), but actually operates to create a still further demand for it on its own account, necessitating the introduction of a further weight, and thus establishing a reciprocal alternation of cause and effect, under the operation of which the very deficiencies themselves are augmented by the means whereby it is sought to repair them. These are objections affecting the principles of the parachute in question, from which those upon the old construction are entirely free. In them the direction of the forces developed in the descent is exactly the most favourable it is possible to conceive, both as regards the retention of the form, and the maintenance of the equilibrium; rendering unnecessary all accessions of weight, save what are required for strength alone, and reducing even those to the smallest possible amount consistent with the actual cohesion of the parts. In the former, on the contrary, the tendency of all these is exactly the reverse; directly opposed to the maintenance of the form, the more they contribute to the retardation of the descent the more they operate towards the destruction of the machine; while their chiefest force being exerted upon the outer edges of the superior surface, should the slightest inequality take place in their action, by which one side becomes operated upon more strongly than another, every thing will favour the derangement of the equipoise, which nothing remains to check but the disposition of the weights themselves. In the present instance, this disposition is the most unfavourable to the exercise of such a restraint that it is possible to imagine. The parachute is stated to weigh 223 pounds, Mr. Cocking 177; it requires but little judgment to foresee how precarious must be the equipoise of a machine so constructed and so disposed. Even the advantage which the remotion of the centre of gravity (which ought to be within the individual himself), would confer, has here been neglected; placed in the very apex of the cone, the slightest inclination will be enough to throw his weight into the body of the parachute, and favour its descent in any way which the deranging circumstances may incline it to assume.

With regard, therefore, to the employment of the parachute in question, or indeed of any other that may be constructed upon the same principle, I have no hesitation in predicting that one of two events must inevitably take place according to the special nature of the defect which may happen to be predominant; either it will come to the ground with a degree of force we have before shown to be incompatible with the final preservation of the individual, or should it be attempted to make it sufficiently light to resist this conclusion, it must give way beneath the undue exercise of the forces it will necessarily develop in the descent.

Besides these essential objections to the projected parachute, there are others of minor importance, chiefly regarding its practical application, but which all taken together, militate greatly against the prospect of its adoption as a convenient mode of regulating a descent. Among these, I shall only mention the difficulty in the first instance of attaching it to the balloon, especially if the wind should happen to be at all high, and the great opposition which it must necessarily offer to the ascent, owing to the permanent state of expansion upon the principle of which it is constructed.

All these disadvantages, the necessary consequences of its shape, are incurred for the sole purpose of avoiding a defect which does not depend upon the shape at all, and which would have been equally avoided by applying the principle of permanent expansion to the usual parachute, or even without any further alteration than by merely increasing the interval between the point of suspension of the individual and the plane of the resisting surface. This would not, it is true, diminish the extent of his deviation from the perpendicular; but by transferring it to a greater distance, it would diminish the angle of oscillation which it subtends, and obviate almost entirely its influence upon the parachute itself.

More might be said on the subject, but that the inutility of the invention does not excuse a further trespass.

M. M.

On the evening of the day following that on which the preceding letter was communicated to the public papers, the melancholy event it had anticipated actually occurred. When at an elevation of about 5,000 feet, the separation was effected, and the parachute commenced descending with a frightful velocity. In a few seconds it was observed to give way before the pressure to which it was subjected in its fall; and in less than two minutes the unfortunate adventurer was lying lifeless on the ground. The average rate at which it thus appears he had descended, was about 40 feet in a second; differing, in fact, only 19 feet from what he would have travelled had the machine remained unbroken. This we are enabled to calculate with greater precision, that we have since been furnished with data upon evidence, whereby the real weight of the apparatus can now be definitively ascertained. It appears from the testimony of Mr. Green, (whose honest practice puts their bungling speculations to shame), that the detachment of the parachute created a difference of about 800 pounds in the actual buoyancy of the balloon. From the same respectable testimony we likewise learn that, of the whole interval traversed by the balloon and parachute in conjunction, the last 3,000 feet were accomplished in the space of five minutes; consequently, that the actual rate of the ascent at the instant of separation was about 36,000 feet (or somewhat under seven miles) an hour. Now the resistance exerted by the atmosphere against a plane surface, in motion at such a rate, is, according to experiments of Mr. Rouse before referred to,[*] exactly in the ratio of the .242 of a pound avoidupois upon each square foot; estimating, therefore, the parachute in its *ascending* form as equivalent to a plane surface of 900 square feet, we shall find that the whole resistance, at the moment of separation, was 900 × .242 = 217 pounds; which deducted from 800 (the number of pounds equivalent to the loss experienced by its detachment), leaves 583 pounds as the actual weight of the entire appendage. Now we have already seen that the parachute in question in its *descending* form, is only equivalent in its powers of resistance to a circular plane of 681 square feet, or about 29 feet in diameter, the terminal velocity of which with such a weight annexed, would have been according to Dr. Hutton's formula $\frac{26}{29}$ $\sqrt{583}$ = 21 feet in a second nearly, or about one-half less than the average rate of its descent when its powers of resistance became weakened by its destruction—perhaps, after all, the greatest blessing which could have occurred; inasmuch as, by the additional impetus it acquired, the fate of the individual may have been consummated at once, and a speedier termination assigned to sufferings which, equally hopeless, might otherwise have been more severe.

[*] For the results of these experiments, see Table, page 303.

APPENDIX C.

THE following list contains the names of all those who
have hitherto essayed the regions of empty air in balloons,
with the dates and places from whence they first ascended,
as far as my present information enables me to determine,
and a slight memoir of such among them as by their
exploits in aerostation, or otherwise, have rendered themselves deserving of especial notice. From this list it will
be seen that the number of aerial voyagers up to the
present period is 471. The proportion in which the
different nations of the globe have contributed to the formation of this catalogue is as follows:—England, 313;
France, 104; Italy, 18; Germany and the German States,
17; Turkey, 5; Prussia, 3; Russia, 2; Poland, 2; Hun-
gary, 2; Denmark, 1; Switzerland, 1; and the States of
North America, 3. Among these are to be found the
names of 49 women; of whom 28 are English; 17 French;
3 Germans, and 1 Italian. Many of those whose names
are here recorded have made frequent repetitions of their
art, while others, and by far the greater number, have no
doubt been limited in their experience to a single occasion.
It would require more research than I can at present
command to determine precisely the aggregate of all

these separate ascents; I have no doubt, however, that they considerably exceed a thousand. Indeed the ascents of Mr. Charles Green alone amount to 249;* those of the different members of the same family to 535, and the whole number executed by Englishmen, as far as I have been able to ascertain, to at least 752.

Out of this vast catalogue of adventurers, the art of aerostation numbers as yet but nine victims in its annals, the fruits of eight unsuccessful experiments. Of these, five, (M.M. Pilâtre de Rosier, Romain, Olivari, Bittorff, and Zambeccari) fell a sacrifice to the inherent perils of the original *montgolfiére*, or fire-balloon. To the fate of one of the remaining four, (Madame Blanchard), the same destructive element, in another guise, likewise contributed; leaving a poor account of three chargeable upon the ordinary casualties of the art, in the form in which it is now practised; and of these three, it will be seen upon reference to the following catalogue,† not one can be fairly said to have been indebted for the issue to circumstances which the exercise of the ordinary precautions might not have enabled him to escape.

* It will be perceived, that since the commencement of the publication of the previous narrative, Mr. Green has added twenty-one leaves to his crown of aeronautical superiority.

* See the memoirs attached to the names of Harris, Mosment, and Sadler (Windham), in their respective places.

A.

ACARD, M. Paris, 4th October, 1802.

ACKERS, Mr. J. Cambridge, 15th May, 1830.

ADAMS, Mr. J. Bath, 16th July, 1824.

ADAMS, Mr. W. H. London, 27th June, 1837.

ADAMS, Mrs. W. H. London, 9th August, 1837.

ADAMS, Mr. J. Jun. London, 29th August, 1837.

ADORN, M. Strasbourg, 15th May, 1784.

ALBAN, M. Javelle, 25th August, 1785.

ALEXANDRE, M.

ALLEN, Captain. London, 20th September, 1837.

ANDREANI, Il Cavalier Don Paolo, along with two brothers of the name of Gerli, ascended at Milan, 25th February, 1784: an exploit which is generally considered to have been the first of the kind ever executed on the other side of the Alps. If, however, any reliance may be placed upon the representations of the European Magazine, an ascent of a most extraordinary description had already taken place at Naples, on the 19th of the same month, in which no less than eight persons participated; of whom three were Italians, three Spaniards, one Frenchman, and one Englishman; the whole party equipped with musical instruments, upon which they continued to perform until they had entirely disappeared from the notice of the spectators.

ANDREOLI, Signor Pasqual. Bologna, 7th October, 1803.

ARBELET, M. d'. Bordeaux, 16th June, 1784.

ARLANDES, Le Marquis d', the companion of M. Pilâtre de Rosier in the first aeronautical excursion from the Bois de Boulogne, Paris, 21st November, 1783. A curious instance of the inconsistent nature of some men's courage; this same gentleman, an officer in the *Garde Royale*, was subsequently broke for cowardice in the execution of his military duties at the commencement of the French revolution.

ARMSTRONG, Mr. J. London, 14th May, 1832.

ARNOLD, Mr. G. Jun. London, 31st August, 1785.

ASTLEY, Mr. W. Ashton-under-Lyne, 9th June, 1827.

B.

BACK, Mr.   London, 6th October, 1836.

BACKHOUSE, Mr.   Ashton-under-Lyne, 5th June, 1827.

BADCOCK, Mrs.   London, 7th July, 1829.

BAILEY, Mr.   Coventry, June, 1828.

BAKER, Mr.   Southampton, 6th September, 1828.

BALDWIN, Mr.   Chester, 8th September, 1785.

BARCLAY, Mr.   London, 27th September, 1836.

BARHAM, Mr.   London, 28th July, 1826.

BARLY, Mr.   Constantinople, 1802.

BARNES, Mr. R. B.   London, 16th June, 1837.

BARZAGO, Signor Giuseppe.

BASS, Mr.   Leith, 27th August, 1830.

BEAUFOY, Captain.   London, 17th June, 1824.

BEAUFOY, Mr. H.   London, 29th August, 1811.

BEAUMONT, Mr. F. W.   Cambridge, 15th May, 1830.

BEAUVAIS, M.   Paris, 14th August, 1798.

BEAZLEY, Mr. S.   London, 11th September, 1837.

BECHET, M.   Rennes, 15th August, 1801.

BECKET, Mr.   London, 17th April, 1827.

BECKET, Miss.   London, 14th June, 1825.

BEER, Mr.   Canterbury, 29th August, 1828.

BEER, Mr. C.   Canterbury, 8th November, 1824.

BERTRAND, M. L'Abbé.   Dijon, 25th April, 1784.

BIGGINS, Mr.   London, 29th June, 1785.

BINN, Mr.   Halifax, 9th August, 1785.

BIOT, M.   Paris, 27th August, 1804.

BISH, Mr.   London, 14th August, 1837.

BITTORFF, Herr, a German aeronaut, who perished at Man-
heim, the 17th July, 1812, in consequence of the accidental
combustion of his balloon.   In his own country he had made
frequent ascents before, with signal success; always, however,
by means of the original *montgolfière* or fire-balloon.   On the
occasion of his death, he had ascended in a balloon of the same
description, and had reached a very considerable elevation when

it took fire, and he was precipitated upon the roofs of some buildings adjoining the outskirts of the town.

BLACKBURN, Miss. Preston, 20th September, 1825.

BLAKESLEY, Captain. London, 21st August, 1837.

BLANCHARD, M. Jean-Pierre, deserves note as having been one of the earliest and most industrious among the votaries of the art of aerial navigation. His first ascent took place from the Champ de Mars, in Paris, on the 2nd March, 1784, upon which occasion it was his intention to have endeavoured to guide the balloon by means of an apparatus which he had constructed for the purpose. In the execution of this design, however, he was disappointed through the impetuosity of a young man, (falsely stated to have been the afterwards celebrated Napoleon, but whose real name was Dupont,) who, unable to obtain permission to share in the honours of the projected experiment, sought to wreak his vengeance upon the balloon, running his sword through the silk and otherwise damaging the machinery by which it was to have been propelled. Restricted therefore to the exercise of the ordinary resources, Blanchard, having hastily repaired the injuries done to his balloon, rose to an altitude of about 9,600 feet, and after having experienced the usual phenomena of aerostation, (particularly interesting at that period of the art, but too well known to sanction repetition here,) descended in safety at Bilancourt, a village near Sèvres, after an excursion of about an hour and a quarter. In the course of his new profession, he visited several of the countries of Europe, England among the rest, and is altogether supposed to have made thirty-six aerial voyages.

BLANCHARD, Mde. Magdeleine-Sophie-Armand, wife of the preceding, and like him for a considerable length of time a successful cultivator of the art to the advancement of which her husband had so much contributed. Her death, which took place on the 7th July, 1819, was occasioned by the ignition of her balloon from a spark of some artificial firework, which she had taken up with her upon the occasion of a nocturnal fête at the gardens of Tivoli, in Paris, whence she had ascended. Her body was dashed to pieces on the parapet of the house, No. 16 Rue de Provence, Faubourg Montmartre.

BLANCHARD, M. Fils. Lubeck, 3rd July, 1792.

BLITZ, Mr. Kidderminster, July, 1829.

BOBY, M. Rouen, 18th July, 1784.

BOLLE, M. Aymè. Scean, 4th August, 1804.

BONAGA, Signor. Bologna, 21st September, 1812.

BOUBERRY, Mr. Coventry, 1828.

BOUCHE, M. Aranjues, 5th June, 1784.

BOXALL, Mr. F. G. London, 9th August, 1837.

BRADLEY, Miss. Warwick, 9th September, 1824.

BRAY, Mr. C. Coventry, 4th June, 1832.

BREMOND, M. Marseilles, 8th May, 1784.

BRIOSCHI, Signor. Padua, 22nd August, 1808.

BROOKE, Mr. Boston, 30th August, 1826.

BROOKES, Mr. Coventry, 25th June, 1824.

BROUGHAM, Miss Anne. Manchester, 4th November, 1837.

BROWN, Mr. London, 5th July, 1802.

BROWN, Mr. Wakefield, 6th October, 1827.

BROWN, Mr. S. Pontefract, September, 1827.

BROWN, Mr. Dewsbury, 1829.

BROWNE, Mr. Manchester, 11th August, 1831.

BRUN, M. Chambery, 12th May, 1784.

BRUNSDON, Mr. Cheltenham, 3rd July, 1837.

BRUNSWICK, Duke of. London, 22nd August, 1836.

BURCHAM, Mr. J. Birmingham, 7th October, 1811.

BURNETT, Mr. London, 21st September, 1836.

BUSHE, Mr. J. London, 23rd June, 1837.

BUTLER, Lord Walter. London, 23rd June, 1837.

BUTLER, Mr. London, 17th September, 1835.

C.

CALLANDER, Major. Alloa, 16th August, 1831.

CAMERON, Mr. Glasgow, 17th April, 1813.

CAMPBELL, Mr. London, 9th August, 1837.

CAPE, Captain. London, 16th July, 1825.

CARNEGIE, Captain.   London, 16th June, 1837.

CARNUS, M. L'Abbé.   Rodez en Guienne, 6th August, 1784.

CARPENTER, Mr.   Manchester, 16th October, 1837.

CARTER, Mr.   Cheltenham, 31st August, 1836.

CARTTAR, Mr. Coroner for Kent. London, 21st August, 1837.

CAVE, Mr. W.   London, 28th July, 1826.

CHADWICK, Mr. J.   Manchester, 4th November, 1837.

CHALFOUR, M.   Bordeaux, 16th June, 1784.

CHAPPE, M. L'Abbé.   Javelle, 17th September, 1785.

CHARLES, M. Professor of Natural Philosophy in Paris, was the first who succeeded in realising the suggestions of Dr. Black and Mr. Cavallo, regarding the applicability of hydrogen gas to the inflation of the balloon ; by which the use of the ancient and dangerous system of ascent by means of rarefied air, has since been entirely superseded.   The first experiment of the kind was made on the 27th August, 1783, with a balloon of thirteen feet in diameter, which being found to succeed, it was resolved that an attempt upon the large scale should immediately be made.   Accordingly, a balloon of twenty-seven feet in diameter was manufactured by MM. Robert, and having been inflated with hydrogen gas which, however, was not accomplished without much labour and expense,) M. Charles and one of the Frères Robert mounted the car, and ascended from the Champ de Mars on the 1st December, 1783.   After a time the balloon returned to the ground, and M. Charles having deposited his companion, reascended alone with enormous velocity, owing to the sudden abstraction of so much weight, and it is supposed reached an elevation of about 10,500 feet.   After having experienced, according to his own description, much inconvenience from the altitude he had attained, he effected his descent without further danger or damage.

CHARTRES, Duc de, afterwards Duke of Orleans, and father of the present King of the French (better known as the unfortunate Egalité), ascended from the park of St. Cloud, 15th July, 1784, in company with MM. Charles, Robert, and another gentleman, whose name I have not been able to ascertain.   In the course of this voyage considerable danger was incurred from the accidental dislodgement of a certain apparatus (disposed within the body of the balloon and designed to regulate its equilibrium by the condensation of atmospheric air,) whereby the action of the valve being obstructed, they were reduced to the necessity of making incisions in the silk to prevent its bursting and enable them to accomplish their descent.

CHASOT, La Comtesse de. Lubeck, 3d July, 1792.

CHATELIER, M. Buchet du. Rennes, 15th August, 1801.

CHEESE, Mrs. London, 4th July, 1836.

CLANRICARDE, Marquis of. London, 2nd August, 1836.

CLARK, Mr. London, 14th August, 1826.

CLARKE, Mr. Wolverhampton, 17th September, 1824.

CLARKE, Mr. London, June, 1836.

CLAUDIUS, Herr. Berlin, 5th May, 1811.

CLAYFIELD, Mr. Bristol, 24th September, 1810.

CLAYTON, Mr. New Orleans, 1836.

COCKING, Mr. Since the commencement of the publication of the previous part of this work, the unfortunate subject of this little notice has ceased to exist, having fallen a victim to an imprudent attempt to descend in a parachute constructed upon defective principles. This melancholy accident is too fresh in the recollection of the reader, and has been too minutely detailed by the public papers to require repetition here. It is due to the memory of the unfortunate, however, to state that he was a person of a most inoffensive disposition, of mild and amiable manners, and though insufficiently acquainted with those branches of science which bore upon the immediate cause of his destruction, not devoid of information upon other matters. He first ascended from Bath or Bristol, I forget which, in company with Mr. Sadler; in his second ascent, which took place from the Vauxhall Gardens, on the 27th of September, 1836, I had the pleasure of accompanying him. His third was the fatal one which terminated his career.

COLDING, Herr. Copenhagen, 5th May, 1811.

COLLET, Mr. London, August, 1836.

COLLIN-HULLIN, M. Paris, 19th September, 1784.

CONST, Mr. London, 14th August, 1837.

COPLING, Captain. London, 22nd September, 1837.

CORNILLOT, M. Pierre. Sevenoaks, 23d August, 1825.

COSSOUL, Madame Virginie. Seville, 29th May, 1823.

COUSTARD-DE-MASSY, M. Nantes, 14th June, 1784.

Cox, Mr. E. London, 21st September, 1836.

CRAMP, Mr. G. Canterbury, 29th August, 1828.

CRAWSHAY, Mr. Norwich, 21st June, 1831.

CRAWSHAY, Mr. Jun. Bury St. Edmunds, 23rd Oct. 1835.

CREAGH, Lt. Col. Sir M. Manchester, 4th November, 1837.

CROMQUELIN, Mr. J. L. London, 16th June, 1837.

CROSBIE, Mr. a gentleman of family and connexions in Ireland, was the first person who ever made public trial of the art in that country. On the 19th January, 1785, he made his first ascent from the Ranelagh Gardens near Dublin, in a balloon of his own construction, when so great was the rapidity of his ascent, owing to the disproportionate allowance of ballast, that he was fairly out of sight in less than three minutes and a half. Opening the valve, however, he shortly after commenced his descent, and finally reached the ground just as he was upon the point of entering upon the sea. On the 12th May, in the same year, Mr. Crosbie announced a repetition of his experiment, with the design of crossing the Channel into England. Being, however, rather a heavy person, and unable from some cause or other to procure a sufficient supply of gas, he had no sooner quitted the ground than he redescended with considerable violence, to the manifest horror of the surrounding spectators; whereupon, a young gentleman from the university, about nineteen years of age, who happened to be present, came forward, and offered to complete the ascent; this being acceded to, Mr. Crosbie resigned his seat, and the honour of the second ascent in that country devolved upon Mr. (subsequently for that same exploit created Sir Richard) Maguire. The second and last attempt of Mr. Crosbie to cross the Irish Channel, was made on the 19th July following. The height to which he ascended on this occasion appears to have been very great, although from the absence of a barometer, he had no means of ascertaining his elevation precisely. The cold, however, was so intense, that his ink had entirely frozen, and the mercury sunk within the bulb of his thermometer. After he had been some time over the sea, the wind veered round to thesouth; the balloon shortly after entered a dense mass of clouds, and becoming charged with moisture, rapidly descended, and in a few seconds became precipitated into the sea. After remaining for a few hours in this perilous situation, drifting before the wind, he was at length overtaken by some vessels that crowded after him, and conveyed, with his balloon, to shore.

CUBIERES, Marquis de. Javelle, 28th August, 1785.

CULLUM, Mr. R. Exeter, 8th September, 1824.

CURRIE, Captain. London, 11th May, 1825.

CUTHBERT, Mr. London, 9th August, 1837.

CUTTILL, Mr. J. B. Lincoln, 29th June, 1830.

D.

DAINTRY, Mr. J. Cambridge, 18th May, 1829.

DAMPIERRE, Marquis de. Lyons, 19th January, 1784.

DAVIES, Mr. R. Hull, 24th September, 1827.

DAVIES, Mrs. London, 1827.

DAVISON, Mr. London, 14th August, 1837.

DAVY, Mr. Beccles, 11th October, 1785.

DAWSON, Miss. Kendal, 30th August, 1825.

DEAN, Mr. London, 23d June, 1825.

DEAN, Miss. London, 9th August, 1837.

DECKER, Mr. Norwich, 1st June, 1785.

DECKER, Mr. Jun. Bristol, 19th April, 1785.

DEGEN, Herr Jacob. Paris, 12th June, 1812.

DELAFIELD, Mr. London, 17th October, 1836.

DELUYNES, M. Nantes, 6th September, 1784.

DERME, Mademoiselle. Paris; 4th August, 1800.

Desgranges, M. Bordeaux, 16th June, 1784.

Devigne, Mr. Constantinople, 1802.

D'Honninctum, Herr. La Haye, 11th July, 1785.

Dickenson, Mr. Stafford, 1830.

Dobney, Mr. J. B. Kidderminster, July, 1829.

Donelly, Mr. J. Bolton, 29th September, 1824.

Dubourt, M. A. dit Augustin. Amsterdam, May, 1808.

Dudley-Stuart, Lord. London, 12th September, 1835.

Dupuis-Delcour, M. a French gentleman, editor of a Parisian journal, and generally connected with literature in that metropolis. In addition to the ascents which he has made in Paris and elsewhere, he has devoted much time to the study of the art, and is in possession of a very complete and interesting collection of tracts and other matters illustrative of its history and practice. I take this opportunity of expressing my obligation to him for the attentions we received from him and his friends upon our sojourn in Paris after the conclusion of our aerial excursion to Germany. His first ascent was from the park of Montjean, near Paris, on the 7th of November, 1824.

E.

Edwards, Miss E. P. London, 8th May, 1827.

Evans, Mrs. London, 24th August, 1836.

F.

Fassy, M. Marseilles, 4th November, 1821.

Finch, Mr. London, 21st August, 1837.

Fitzpatrick, Colonel. Oxford, 29th August, 1785.

Fleurand, M. Lyons, 4th June, 1784.

Fontaine, M. Lyons, 19th January, 1784.

Forster, Dr. T. Chelmsford, 30th April, 1831.

Forte, Mr. Cheltenham, 22nd September, 1836.

Fox, Mr. G. London, 17th April, 1827.

Foxcroft, Mr. Lancaster, 30th July, 1832.

French, Lieutenant. Chester, 1st September, 1785.

Frobisher, Mr. Halifax, 9th August, 1785.

G.

GAERTNER, Herr.   Berlin, 13th April, 1803.

GAMBLE, Mr.   London, 15th September, 1837.

GANDY, Lieutenant.   Portsea, 25th October, 1824.

GARDINER, Mr. W. H.   Lewes, 30th September, 1828.

GARNERIN, M. André-Jacques, a French aeronaut, noted for
the number and boldness of his exploits. Of these the most remarkable were the night ascents, the first of which took place
from the Tivoli gardens, at Paris, on the 4th August, 1807 ; the
balloon being decorated with lamps, and charged with artificial
fireworks intended to be ignited when at a sufficient elevation
in the air.   Upon a repetition of this experiment from the same
place, on the 21st September following, he appears to have incurred considerable danger from the extreme distension of the
balloon, owing to the inability under which he lay of discharging
the gas for fear of its taking fire at some of the numerous lamps,
which yet from the distance at which they were disposed, he was
unable effectually to extinguish.   In these straits he was fain to
continue until the following morning, when the lights expiring
of their own accord, permitted him to take the necessary steps
for his descent.   The name of Garnerin is, however, more particularly remembered on account of his having been the first
who ever safely descended from a balloon by means of a parachute ; an exploit which he originally performed in an ascent
from Paris, on the 21st of October, 1797, and afterwards repeated on various occasions in England, France, and other
places upon the Continent.   The number of his ascents is said
to have exceeded fifty.

GARNERIN, Madame Jacques, (under the name of Labrosse)
Paris, 10th November, 1798

GARNERIN, Mademoiselle Elisa. Paris, 2nd Sept. 1815.

GARNERIN, Mademoiselle Eugene.

GARNERIN, Mdlle. Cécile-Benoit. Paris, 26th April, 1818.

GARNERIN, Mademoiselle Blanche. Paris, 26th April, 1818.

GARRETT, Mr. Cheltenham, 22nd September, 1837.

GAY-LUSSAC, M., the celebrated French chemist, ascended in company with M. Biot, another French philosopher of like celebrity, from the Conservatoire des Arts, in Paris, August 23rd, 1804, with a view to instituting experiments in various branches of the physical sciences. At an elevation of about 6,500 feet above the level of the sea they commenced their observations, from which, however, at that altitude they obtained no results of sufficient importance to merit attention; unless, indeed, it can be considered as such that no difference was discernible in the action of the different natural powers of magnetism, electricity, and galvanism, as far as they were tested, from what they would have displayed upon the surface of the earth. At an elevation of about 11,000 feet they let go one of the birds with which they had been provided; for a moment it rested upon the edge of the car; then, launching into the deep abyss, directed its course in gradually extended circles towards the earth; thereby refuting an erroneous opinion generally prevalent, that the rarity of the atmosphere in such elevated situations was incompatible with the exercise of their natural functions.

The humidity of the surrounding medium, always unfavorable to the display of electrical phenomena, seems to have operated to a considerable extent on the present occasion, as I can only find one conclusion in that branch of science to which they were able to arrive, that is worthy of being recorded. Upon letting down a wire of about 250 feet in length they obtained by means of the electrophorus indications of negative electricity from its upper extremity; a result which appears strongly to confirm the opinion given by Volta and M. Saussure regarding the increase of the electric matter in the superior regions of the sky. Having obtained an elevation of about 13,000 feet without being able to discover any thing further meriting to be related they commenced their descent, and at half-past one alighted near the village of Merivale, in the Department of the Loiret, about fifty miles from Paris. At their greatest elevation the thermometer stood at 30° of Fahrenheit, or two degrees below the freezing point of water.

The results of this experiment not having realised the expectations of the scientific world in Paris, owing chiefly to the limitation which the weight they carried imposed upon their ascent, it was resolved to make another attempt, in which M. Gay-Lussac should go up alone; accordingly on the 15th of September, in the same year, that gentleman again ascended from the same place, with the intention of prosecuting his enquiries as far as it was possible without compromising his own safety and the success of the undertaking in which he was embarked. The results of this experiment are consequently much more important to science than those of the last, and deserve a notice much more detailed than the narrow limits which we have prescribed to ourselves will permit us to bestow. At an elevation of 9,929 feet the oscillations of the horizontal needle appear to have been considerably accelerated; giving a result of 20 in 83″, although on earth the same number would have required 84 . 33″. The variation as far as his means would enable him to determine, appeared at the height of 12,651 feet to be exactly the same as below; his conclusions, however, upon this point must still from the nature of the experiment, be considered as liable to much uncertainty. With regard to the dip, the derangement of the instrument, occasioned by the great drought, prevented him from making any observations.

The hygrometrical condition of the atmosphere throughout the whole voyage, as observed at different altitudes, presents an appearance of irregularity which at first sight would lead to a suspicion that accident or local circumstances alone had any share in determining its variations. Upon referring, however, to the tables at the end of the narrative, as published in the "Annales de Chimie," and comparing these changes with those indicated in the thermometrical column at the same altitudes, it will be seen that the actual humidity of the atmosphere was by no means irregular in its gradations, but on the contrary, followed a very certain though rapidly decreasing series.

At the height of about 21,500 feet, two large glass balloons, which had been previously exhausted of air, were opened for the purpose of subjecting to analysis a portion of the atmosphere abstracted from such an elevation. The result of this enquiry, although not arrived at till after the conclusion of the voyage, it may be as well to observe here, proved no difference whatever in the composition of the atmospheric volume, and strongly confirms the opinion that, unless when contaminated by the accidental introduction of foreign matter, the nature and proportion of its constituents are invariably the same.

Having reached a height of 22,977 feet above the level of the sea, and almost all his ballast having been already disposed of, he considered it prudent to arrest his farther progress. The appearance of the sky at this, the extreme point of his ascent, was particularly interesting; the colour, especially about the zenith, being comparable only to a fine shade of Prussian blue, while notwithstanding his excessive elevation, he could still perceive the clouds in considerable abundance floating at an apparently immeasurable distance above his head. The temperature had fallen as low as 17.1 of Fahrenheit. Although well clothed, he now began to suffer from the effects of the cold, especially in his hands, which he was obliged to keep exposed, and constantly employed in handling the various instruments necessary in making observations. His respiration likewise, he describes as being sensibly affected, and, as well as his pulse, considerably accelerated; not so much, however, as to occasion him any great inconvenience, or to render it necessary for him to precipitate his descent. Owing to the increase in the rate at which the former function was performed, and the siccity of the atmosphere upon which it had to act, it is not surprising that his throat should have become so dry that it was with difficulty he was able to swallow a few mouthfuls of bread. These were the only physical inconveniences which he experienced, and if we consider all the collateral circumstances of the case, the state of his health at the time, the great fatigue he had undergone, the want of sleep throughout the previous night, and the anxiety and interest (to speak in the most moderate terms) which he may be supposed to have felt in some degree regarding the issue of such a venture, I think there will not remain much to charge to the account of his excessive elevation. At a quarter to four he came to the ground, and effected his landing near the small hamlet of St. Gourgon, within about six leagues of Rouen, having been altogether five hours and three-quarters in the air.

I have dwelt somewhat more largely upon the preceding, as they are in fact the only ascents hitherto executed by which the interests of science have been in any way advanced.

GEE, Mr. Stockport, 18th June, 1827.

GEE, Mr. R. Ashton-under-Lyne, 9th June, 1827.

GIARD, M. Florence, 1st October, 1811.

GERLI, Signor Carlo Giuseppe. Milan, 25th February, 1784.

GERLI, Signor Augustino. Milan, 25th February, 1784.

GLASFORD, Mr. London, 3rd August, 1802.

GOCHER, Mr. Bury St. Edmonds, 23rd October, 1835.

GOWARD, Mr. Ipswich, 1827.

GOWARD, Mr. Junior. Ipswich, 1827.

GRAFTON, Mr. Macclesfield, 25th June, 1827.

GRAHAM, Mr. London, 5th September, 1823.

GRAHAM, Mrs. London, 2nd June, 1824

GRASSETTI, Dr. Bologna, 7th October, 1803.

GREEN, Mr. Thomas. Mansfield, April, 1827.

GREEN, Mr. Charles. London, 19th July, 1821.

GREEN, Mr. George. London, June, 1825.

GREEN, Mr. George, Junior. London, 30th March, 1826.

GREEN, Mr. Henry. London, 30th March, 1826.

GREEN, Mr. James. London, 1827.

GREEN, Mr. William. Newcastle-upon-Tyne, Sept. 1825.

GREEN, Mrs. Charles. London, 23rd July, 1836.

GREEN, Mrs. Henry. Rochester, 31st May, 1828.

GREEN, Miss Marianne. London, 9th September, 1836.

GREETHAM, Lieutenant. Portsea, 15th July, 1829.

GREGG, The Rev. G. Belfast.

GRIFFITHS, Mr. Cheltenham, 1823.

GRISOLLE, M.

GRONOW, Captain. Paris, 19th December, 1836.

GUILLIE, M. Charles.

GYE, Mr. Junior. London, 17th October, 1836.

GYE, Mr. E. London, 9th September, 1836.

H

HABRO, Mr. Worcester, 20th September, 1824.

HARPER, Mr. Birmingham, 4th January, 1785.

HARRIS, Lieutenant, an officer in the British navy, perished in an ascent which he effected in company with a young lady, Miss Stocks, from the Eagle Tavern, City Road, London, the 25th May, 1824. His death was entirely attributable to his own imprudence in tampering with the details of an art in which he was not sufficiently versed. Among the other alterations which he had inconsiderately thought proper to make in the mechanical construction of the balloon was the inordinate enlargement of the valve for the purpose of facilitating the escape of the gas when the voyage should have been terminated. As the use of so large an aperture while in the air would be attended with much risk, another valve of the ordinary size and for the ordinary purposes, was contrived in one of its flaps or doors; an arrangement which necessitated the use of two valve lines, one for each of the two separate openings.

With a balloon of this construction Mr. Harris ascended on the day above-mentioned. Previously to his departure, to avoid the possibility of his confounding the two lines during the course of his excursion in the presence of several of his friends (who, however, were themselves equally unacquainted with the common phenomena of practical aerostation), he proceeded to make that which was attached to the larger aperture a fixture to the side of the car, leaving the other at liberty to be used as occasion might require, and thus commenced his ascent. This fatal error was the cause of his destruction. No sooner had the luckless aeronaut prepared to terminate his excursion, and for that purpose commenced a discharge of gas through the proper opening, than a longitudinal extension of the apparatus (the natural consequence of its diminished contents), immediately ensued, and a strain amounting to the whole weight of the car became thrown upon the line which unfortunately connected it with the doors of the preposterous aperture above. The consequences are evident; the valve was opened, the gas rushed through in torrents, the balloon descended with frightful rapidity, and the cause increasing with the effect, long before it reached the earth became nearly stripped of its contents. In a few seconds they were dashed to the ground, where they were shortly after discovered, buried beneath a monumental pile of silk and net-work; fit emblems of the fate which his own rash act had drawn upon him. Mr. Harris was destroyed upon the spot. His companion, Miss Stocks, was somewhat more fortunate. In the confusion which followed the first announcement of their perilous situation she had fainted, and fallen forward, it is supposed, upon the body of Mr. Harris, who was thus doomed to sustain the first violence of the shock. Another providential circumstance likewise contributed to her preservation; a large branch of a tree, the only one for miles around suitable to such a purpose, projected horizontally, and intercepting the car as it descended, served considerably to mitigate the extreme force of the concussion.

I have been led to enter somewhat more minutely into the details of this disasterous case than I should otherwise perhaps have felt authorized, from a consideration of the mystery which has hitherto clouded the event, and the false and scurrilous reports to which it has given rise. That the above however, (for which I confess myself indebted to Mr. Green), is the true explanation of this melancholy occurrence, presumption amounting to proof, is afforded in the facts, first, that when the balloon was examined no rent or damage could be detected through which the gas could have accidently escaped, while in the second place, Miss Stocks declares that she distinctly heard the peculiar sound which always accompanies the shutting of the valve, as soon as Mr. Harris had let go the line; a circumstance which proves that the fatal issue of gas could not have taken place through any deficiency in the construction of the legitimate outlet. How far Miss Stocks was satisfied that the occurrence was attributable to the inexperience of her unfortunate comrade, and not to the inherent perils of the art, appears from the fact that she has since accompanied Mr. Green in three several ascents.

HARRISON, Miss.   London, 6th October, 1836.

HARVEY, Colonel.   Norwich, 7th September, 1825.

HEMMING, Mr.   London, May, 1828.

HENRY, Mademoiselle Celestine.   Paris, 28th August, 1798.

HILDYARD, Mr. R. C.   Lancaster, 16th August, 1832.

HILL, Mr.   Hull, 24th September, 1827.

HODGES, Mr.   London, 16th August, 1836.

HODGKINS, Mr.   Doncaster, 4th September, 1827.

HOLLOND, Mr. Richard.   London, 1836.

HOLLOND, Mr. Robert.   Cambridge, 15th May, 1830.

HOPE, Mr. J.   Cambridge, 14th May, 1832.

HUGHES, Mr. R.   Paris, 19th December, 1836.

HUGHES, Mr. W.   London, 9th September, 1836.

HUGHES, Mr. T.   London, 21st September, 1836.

HULKES, Mr.   Cambridge, 15th May, 1830.

HUME, Mr. J.   London, 16th May, 1837.

I

IBRAIM PACHA.

J

JEFFERYS, Mr. Birmingham, 1829.

JEFFRIES, Dr. an American gentleman, first ascended from London on the 30th November, 1784, in company with M. Blanchard, with whom he afterwards crossed the channel from Dover to Calais, on the 7th January in the following year. This undertaking was not accomplished without considerable difficulty. Owing to the condensation which ensued while over the sea the balloon began to descend, obliging the aeronauts to throw over every thing they possessed, even to their very habiliments, in order to arrest its depression, until they should have made the opposite coast. Just, however, when they appeared to be in the greatest straits, and had already rejected every thing they could possibly dispense with, the balloon of its own accord began to reascend, and carrying them clear of all danger, deposited them in safety in the forest of Guiennes, where a small monument appears to signalize the place of their descent.

JEPSON, Mr. London, 1832.

JONES, Mr. Hereford, 1829.

JUNGIUS, Professor. Berlin, 1805.

K

KENNEDY, Captain. Gainsborough, 10th September, 1827.

KENNET, Miss H. Chelmsford, 1831.

KENNET, Miss E. Chelmsford, 1831.

KENT, Dr. London, 1836.

KUPARENTO, M. Jordaki, a Polish aeronaut, ascended from Warsaw, 24th July, 1808, in a *montgolfière*. When at a considerable elevation his balloon took fire, but being provided with a parachute he was enabled to descend in safety.

L

LALANDE, M. de. Paris, 1799.

LAPORTE D'ANGLEFORT, Le Comte de. Lyons, 9th Jan. 1784.

LAURENCIN, Le Comte de.   Lyons, 9th January, 1784.

LEEDS, Mr. W.   Cambridge, 14th May, 1832.

LEICESTER, Captain.   London, 23d June, 1837.

LEIGH, Captain.   Warrington, 26th July, 1827.

LEMERCIER, M.   Paris, 19th October, 1805.

L'EPINARD, Le Chevalier de.   Lisle, August, 1785.

LEWIN, Mr.   London, 16th June, 1837.

L'HOEST, Herr.   Hamburgh, July, 1803.

LIGNES, Le Prince Charles de.   Lyons, 9th January, 1784.

LIVINGSTONE, Mr.   Dublin, 1821.

LOCKER, Mr.   London, 4th July, 1802.

LOCKWOOD, Mr.   London, 3rd June, 1785.

LOUCHET; M.   Rodez en Guienne, 6th August, 1784.

LOYED, Mr.   Birmingham, 1829.

LUNARDI, Signor Vincenzo, a gentleman attached to the Neapolitan Embassy in London, whose ascent from the Royal Artillery grounds, Moorfields, on the 15th September, 1784, is generally considered to have been the first experiment of the kind ever instituted in these realms.   This assertion, however, is incorrect; Mr. Tytler, of Edinburgh, having already ascended from that city on the 27th of the preceding month.   The exploits of Signor Lunardi were not, however, confined to one occasion only, or to one portion of the British dominions.   In the month of November, in the following year, he made an ascent from Heriot's Hospital Gardens, in Edinburgh, being the second aerial excursion ever witnessed north of the Tweed. This experiment he repeated again from the same place, in the following month; after which he ascended from Glasgow, Kelso, and other places in Great Britain, and finally returned to Italy, where he continued for some time to gratify his countrymen with frequent repetitions of his adventurous exploits.

LUWOP, Le General.

LUZARCHES, Madame.

LYSTER. Colonel.   Maidstone, 1828.

M

MAGUIRE, Sir Richard, ascended from the Royal Barracks, Dublin, 12th May, 1785, and was the second person in that country who hazarded the experiment. The circumstances of his ascent are not without interest. Mr. Crosbie, (a slight account of whose exploits will be found in the notice attached to his name in the present Appendix), having on the occasion in question made an unsuccessful effort to quit the ground, owing to the insufficient inflation of his balloon, the hero of this little notice, then a young man at the university who happened by accident to be present, stepped forward and volunteered to complete the ascent; his offer being accepted he leapt into the car and changing places with Mr. Crosbie, immediately and rapidly rose to a considerable elevation. Being carried by the wind in the direction of the Irish channel, he was however shortly forced to descend; an operation which indeed he had just time to accomplish before he reached the shore. Immediately upon his landing he was seized by the crowds who had followed to assist him, and borne in triumph upon their shoulders into town. Shortly after he was knighted for his adventure by the Lord Lieutenant, and if I mistake not is at the present moment living in the enjoyment of his honours.

MAISON, Mademoiselle.

MAITRE, Le Chevalier.

MALCOLM, Mr. Salford. 1826.

MALTITZ, Baron. London, 16th May, 1837.

MARCHESELLI, Signor. Genoa, 10th November, 1811.

MARET, M. Marseilles, 8th May, 1784.

MARGAT, M.

MARGAT, Madame. Paris, 4th June, 1818.

MARR, Mr. London, 17th May, 1837.

MARSHALL, Mr. Worksop, 1833.

MARSHALL, Mr. Derby, 1828.

MARSHALL, M. J.   Norwich, 28th June, 1831.

MATTHEW, Captain, R. N.   Hereford, 1828.

MERRY, Mr.   London, 1832.

MICHAUX, M.

MILLER, Mr.   Peterborough, 1831.

MILNES, Mr.   Cambridge, 1828.

MONEY, Major, first ascended from London, in company with Sir Edward Vernon, Mr. Lockwood, and Mr. Blake, (whose name I find unaccountably omitted from its proper place), on the 3rd June, 1785. In a subsequent ascent from Norwich, on the 22nd July in the same year, he had the misfortune to be blown out to sea, into which, after having continued for two hours hopelessly hovering over it, he was finally compelled to descend. Here he remained for seven hours struggling with his fate, until at length, just as he had abandoned all hope of succour, he was rescued by a revenue cutter which had been despatched to his assistance.

MONRO, Mr. W.   London, 31st July, 1837.

MONTGOLFIER, M. Joseph, one of the two brothers to whom is ascribed the honor of having first demonstrated the practicability of the views of Dr. Black and Mr. Cavallo regarding the elevation into the atmosphere of solid bodies by attaching them to vessels filled with materials specifically lighter than the surrounding air. To what extent the subject of the present notice may have participated in the original discovery of the balloon is not easy to determine; from what I can collect, however, I am rather inclined to think that his share, if any, was extremely small, and that had it not been for the interest he afterwards took in its practical application as an aeronaut (in which character alone he claims our consideration here) his name would never have been coupled with that of Etienne, to the detraction of the latter as the real inventor of the art of aerostation.

The principal claim which Joseph Montgolfier has upon our notice is in connexion with an ascent which took place at Lyons, 19th January, 1784, and which, for the magnitude of the scale upon which it was conducted exceeds all that has hitherto been effected in the same line.

The balloon which was employed upon this occasion was a pyriform vessel, constructed of two layers of linen cloth, enclosing one of paper between them, (for the purpose of increasing its imperviousness), and measured when duly inflated 130 feet in height, and 105 in breadth; it was capable of containing 540,000 cubic feet of air, and when charged for the ascent, supported with ease seven persons, and ballast to the amount of 3,200 pounds, independant of its various accessories,— its car in the form of a gallery, 72 feet in circumference, accommodated with seats, four feet wide and eight apart; its furnace, 20 feet in diameter, with its proportionate store of fuel made up into faggots of wood and straw; its massive frame-work to maintain the inferior aperture; its drapery, netting, cordage, implements and other requisites all in the same proportion, the weight of which it would not be easy at present to determine. The names of those who participated in the honour of this expedition, were Joseph Montgolfier himself, under whose direction the whole had been got up, Pilâtre de Rosier, le Comte de Laurencin, le Marquis de Dampierre, le Comte D'Anglefort, le Prince Charles de Lignes, and a young man named Fontaine, who happening to be in the car at the moment, when, suddenly lightened by the hasty departure of another gentleman, it escaped into the air, became accidentally included in the party. In a few seconds it rose to an elevation of about 3,000 feet; an opening, however, of about four feet in length which appeared above the equator of the balloon, soon brought it down again, with a velocity even greater than its ascent, and it reached the ground at a distance of about 1,200 feet from the place of its departure.

Joseph Montgolfier, the subject of the above little notice, was the only one of the family who appears to have carried the designs of his more scientific brother into actual execution. It is not generally so understood, but Etienne Montgolfier, the original discoverer of the art, never himself ascended; at least so as to come before the public in the character of a practical aeronaut, and indeed I am not certain that Joseph himself ever repeated the experiment.

MORVEAU, M. Guyton. Dijon, 25th April, 1784.

MOSMENT, M., a French aeronaut, made his last ascent at Lille, the 7th April, 1806. This gentleman was in the habit of ascending upon a platform, which served him instead of a car; a method to which he adhered upon the present occasion. About ten minutes after he had quitted the earth, a small parachute, containing some animal, was observed to be launched from the car; immediately afterwards an object, which was soon ascertained to be the flag of the adventurer, was perceived slowly following it through the sky. A rumour now began to be spread about, that the aeronaut himself had been precipitated from the car, which was shortly after confirmed by the discovery of his body, almost buried under the sand in one of the fosses of the ramparts that surround the town. It is supposed that the oscillations communicated to the balloon in the act of delivering the parachute had thrown him off his balance, which the absence of the usual bulwark prevented him from recovering. Some persons, indeed, pretended at the time to have heard him declare before-hand the event, and from thence argue that the affair was not unpremeditated, although from what design is not apparent.

MOUCHET, M. Nantes, 14th June, 1784.

N

NANCY, Mademoiselle.
NARBONNE, Le Comte de.
NOLLIN, M.

O

OGLE, Captain. London, 6th October, 1836.

OLIVARI, Signor. The death of this gentleman occurred at Orleans, on the 25th November, 1802, in an ascent which he made at that place in a *montgolfière* or fire balloon, constructed simply of paper, held together only by a few strips or bands of linen. His car, which was made of wicker-work, and suspended directly beneath the furnace, being charged with materials destined for its supply, became at a great height the prey of the devouring element, and the unfortunate aeronaut was precipitated to the earth at about a league from the place of his departure.

ORLANDI, Signor. Bologna, 1828.

OYESTON, Miss. Newcastle-upon-Tyne.

P

PAGET, Lieutenant. London, 12th August, 1811.

PATRICK, Mr. London, August, 1836.

PAULY, M. Paris, 19th October, 1805.

PHILLIPS, Mr. H. L. Manchester, 3rd August, 1827.

PICKERING, Mr. Chichester, 14th October, 1828.

PILTAY, M. Paris, 19th December, 1836.

PIRI, Signor. Paris, 9th January, 1837.

POLHILL, Captain. London, 15th May, 1837.

POTAIN, Dr. Dublin, 17th June, 1785.*

---

* In a little brochure which I published in Paris upon the occasion of the aeronautical expedition to Weilburg, I was erroneously led to cast some doubts upon the truth of the accounts given of this ascent. The fact is, that from some neglect, or other reasons which it would be impossible at present to fathom, no notice appears to have been ever taken of this voyage; a circumstance the more remarkable as having happened in Ireland, where such experiments have, comparatively speaking, been few, and at a period when as yet they were by no means common any where. Such however is the case; nor, although several persons recollected the ascent as having occurred, when it was once recalled to their notice, did I ever meet with any one who was acquainted with the circumstances or name of the individual. Convinced of my error, I am

Porosky, Le Comte J. de.

Power, Mr. T. London, 23rd June, 1837.

Price, Mr.

Proust, Dr. J. Versailles, 23rd June, 1784.

Puckle, Mr. Lincoln, 1831.

Pugh, Mr.

Puymarin, M.

R

Radcliffe, Mr. C. Blackburn, October, 1825.

Rambaut, M. Aix, 31st May, 1784.

Ramshay, Mr. Carlisle, 12th October, 1825.

Reichard, M.

Reichard, Madame Vilhemine. Brussels, 22nd Nov. 1818.

Reynolds, Mr. J. London, 16th June, 1837.

Richard, M. Jean-Marie. Montjean, 7th November, 1824.

Richardson, Mr. J. Derby, 1828.

Rivierre, M.

Robert, M. Aîné. Paris, 1st December, 1783.

Robert, M. Cadet. Paris, 19th September, 1784. These
two brothers were instrumental with M. Charles in the construction of the first hydrogen-gas balloon, in contradistinction
to those previously used, in which the agent of the ascension
was atmospheric air, rarefied by artificial heat.

Robertson, M. Etienne Gaspard. Hamburgh, July, 1803.

Robertson, Madame.

Robertson, M. Eugene. Lisbon, 12th December, 1820.

Robertson, M. Dimitri. Calcutta, 1836.

Robinson, Mrs. Canterbury, 2nd September, 1828.

Roger, Mr. Kilmarnock, 1830.

extremely glad to have this opportunity of doing justice to the character
of a most respectable veteran of aerostation, who is at this moment
living in Paris, and whose feelings, by an unjust suspicion, I fear I
have most unhappily wounded.

Rolens, Mr. Rochester, 1828.

Romain, M., the companion of M. Pilâtre de Rosier in the disastrous ascent from Boulogne which fatally terminated the career of both. For the particulars see the account given under the head of the latter.

Roscoe, Mrs. Paris, 19th December, 1836.

Rosier, M. François Pilâtre de, was the companion of the Marquis d'Arlandes in the first aeronautical expedition ever attempted. It would lead us into a regular history of the rise and progress of aerostation, and far exceed the limits allotted to the present appendix, were I to attempt to trace with any thing like the precision they deserve, the proceedings of this remarkable man as far as they are connected with the present subject. It appears from all accounts that he must have been a person of a very bold and adventurous character, of a most enthusiastic disposition, and possessed of as much scientific education as the age in general could boast of. He first came into note for his experiments upon the recovery of persons apparently drowned, in the course of which he had ample opportunity for the development of those qualities and acquirements by which he afterwards rose to higher fame. Upon the occasion of a great aeronautical experiment which was about to take place in the presence of the court of France, he strenuously exerted himself to obtain permission to accompany the Marquis d'Arlandes, who had already declared his intention to occupy the car, and thus become one of the two who first dared to trust themselves to the conduct of the winds in the then dangerous conveyance of the fire-balloon. This event took place from the palace of La Muette, in the Bois de Boulogne, near Paris, on the 21st November, 1783. From this time forward, until the fatal termination of his career, M. Pilâtre de Rosier seems entirely to have devoted himself to the practice and improvement of the art of aerostation; a a pursuit in which, however, he was not long destined to continue. On the 15th June, 1785, in company with a young gentleman named Romain, he ascended at Boulogne sur Mer, with the intention of reversing the experiment lately performed by M. Blanchard and Dr. Jeffries, and crossing the channel into England by means of his balloon. Unfortunately the arrangements which he adopted to secure his success were the cause of his failure as well as of his destruction. In order to counteract the fluctuations consequent upon all aerial excursions under the ordinary circumstances, and obtain the power of increasing or diminishing the weight of his apparatus at will, without the usual expenditure of gas and ballast, he had conceived the idea of uniting in one the two systems of MM. Montgolfier and Charles, and accordingly affixed to the hydrogen-gas balloon, by which the principal part of the weight was to be borne, a small *montgolfière* or fire-balloon, by acting upon which he expected to be able to alter his specific gravity as occasion might require. The theory was correct; the error lay in the application. Distended in the course of its elevation, the inflammable contents of the larger sphere soon filled the vacant portions of the silk, and pouring down the tube which formed the neck of the balloon speedily reached the furnace, which was disposed at its lower extremity, and became ignited. The whole apparatus was consumed in the air, and the two unfortunate voyagers precipitated upon the rocks which bound the shores of the sea between Calais and Boulogne. A monument has since been erected on the spot where they fell, which by a singular coincidence is not far from that which now stands to perpetuate and attest the successful exploit of Blanchard and Dr. Jeffries.

Rossi, Signor Gaetano.

Rossiter, Mr. London, 1st July, 1824.

Rouse, Mr. London, 15th May, 1837.

Rousseau, M. Navan, 14th April, 1784.

Routh, Dr. Beccles, 11th October, 1785.

Ruggieri, Signor Claudio. Paris, 9th November, 1801.

Rush, Mr. London, 21st September, 1836.

Russon, Mr. Leeds, 18th August, 1830.

T

S

SADLER, Mr. James, one of the earliest of those who in this country applied themselves to the practice of aerostation, made his first ascent from Oxford on the 12th of October, 1784. The correctness of this account has, it is true, been called in question by Mr. Tiberius Cavallo, in his history of the early stages of the art, upon the grounds of insufficient testimony, and the event in question referred to the 12th of the November following, when that which would otherwise have been his second ascent is universally admitted to have taken place. On the other hand again, it is but just to observe that notices of the preceding are to be found in several of the periodical publications of the day, and in almost all the historical essays upon aerostation in the various encyclopædias compiled from the same sources; in addition to which we have to add the testimony of his son, who in a letter in reply to my enquiries on the subject, expressly alleges his father's first ascent to have taken place at Oxford, on the 12th of October in the above year, from the gardens of the botanical establishment belonging to that University. Admitting its authenticity, however, this was not the first occasion upon which Mr. Sadler appeared a candidate for aeronautical honours; having on the 12th of the preceding September (about three days before the experiment of Signor Lunardi from Moorfields) made an ineffectual attempt to ascend in a *montgolfiére*, from a retired spot in the neighbourhood of Shotover Hill near Oxford, which was frustrated by the accidental combustion of the balloon almost immediately after it had quitted the earth and before it had attained an elevation of twenty yards. Had it not been for this untoward accident a foreigner would not have had to boast the honour of having accomplished the first aerial voyage ever executed in England. From this period, for a considerable number of years, Mr. Sadler continued to enjoy the reputation of one of the ablest and most intrepid aeronauts of the age; a character which he supported by numerous brilliant and well-conducted experiments in different parts of the United Kingdom. In one of these, which came off from the Belvidere Gardens in Dublin, on the 1st October, 1812, he made an attempt to cross the Irish Channel, in which he had very nearly succeeded, when he was forced by adverse winds to descend into the sea off Liverpool, in order to avoid being carried still further from the coast. After having remained for a considerable length of time in this perilous situation he was rescued by a fishing-boat that happened luckily to witness his fall. A similar accident had befallen him not long previously, in company with Mr. Clayfield, off Bristol, on the 24th September, 1810. The precise number of his ascents I have not been able satisfactorily to determine, but I have reason to believe they do not much fall short of sixty. Of his two sons, John and Windham, who both followed in the same career, the former is still living, although he has for some time discontinued the practise of the art, after having for many years worthily supported his father's reputation. The melancholy fate of the latter is still fresh in the recollection of the public.

SADLER, Mr. John. Worcester, 13th May, 1785.

SADLER, Mr. Windham, the second son of the veteran aeronaut of that name, was the last who fell a sacrifice to the practice of aerostation. On the day of the ascent the weather was extremely rough; so much so indeed, that a gentleman who was destined to be his companion refused to proceed with him, and his place was taken by Mr. J. Donelly, a servant of the unfortunate aeronaut who happened to be present. At the conclusion of the voyage, and after a considerable quantity of gas had been allowed to escape, by some neglect or other on the part of the bystanders, the rope by which the balloon had been retained was suddenly let go. The balloon, borne from the ground more indeed by the force of the wind than its own ascensive power, immediately sprung up, and sweeping violently along, brought the car into contact with the chimney of a house which stood in its progress, and thus it is supposed, by the force of the blow deprived Mr. Sadler of the power of exerting himself for his preservation. At any rate, whether from the loss of sense, or of equilibrium from the violence of the concussion, the luckless aeronaut fell over the side of the car, and after hanging for a few seconds suspended by the feet, was finally precipitated on the ground, near the town of Blackburne, on the 29th September, 1824. In one of his excursions, July 22nd, 1817, Mr. Sadler, more fortunate than his father, cleared the Irish Channel from Dublin to Holyhead.

SAGE, Mrs. in company with Mr. Biggins, ascended from London, 29th June, 1785, on which account she claims the distinction of having been the first Englishwoman that ever ventured to explore the regions of empty space in a balloon.

SAINTE-CROIX, M. Salisbury, 10th August, 1786.

SAUNDERS, Mr. B. Bristol, 29th July, 1824.

SAYWELL, Mr. Nottingham, 3rd August, 1826.

SCOTT, Mr. G. W. Cambridge, October, 1829.

SEGUIN, M. A. de la Salle. Paris, 9th January, 1837.

SELIM-OGAT. Smyrna, 1825.

SERGEANT, Mr. Stamford, 11th September, 1826.

SHELDON, Mr. London, 16th October, 1784. This gentleman is generally stated to have been the first native of Great Britain that ever made trial of the aerial car. This position is, however, incorrect; both Mr. Tytler, in Edinburgh, and (there is every reason to believe) Mr. Sadler, in Oxford, having already had precedence, the former on the 27th of August, and the latter on the 12th of October in the same year.

SIMMONDS, Mr. H. Reading, 1st August, 1823.

SIMON, M. London, 21st August, 1837.

SIMONET, Mademoiselle. London, 3d of May, 1785. This young lady, who was but fourteen years and a half old when she accompanied M. Blanchard upon the above occasion, is entitled to notice as having been the first female that ascended in Great Britain, and the second, I believe, by whom the experiment was hazarded in any part of the world.

Simonet, Mademoiselle, jeune. London, 21st May, 1785.

Simpson, Mr. Stamford, 11th September, 1826.

Simpson, Mr. Mansfield, September, 1829.

Slea, Mr. Brighton, 8th October, 1824.

Slea, Mr. C. Penrith, 13th June, 1832.

Sloan, Mr. Manchester, 4th November, 1837.

Smith, Mr. W. Birmingham, 31st August, 1827.

Sneath, Mr. ascended from Bleak Hill, near Mansfield, on the night of the 24th of May, 1837, in a fire-balloon of his own construction, being the only instance that I am aware of, with the exception of Mr. Tytler's from Edinburgh, in which such an expedient has succeeded in any part of the British dominions. After remaining in the air for two hours the balloon began to descend, and at eleven the grapnel took effect in a hedge near the village of Spondon. Apprehensive of the escape of the balloon in case he should quit it, and afraid to allow the fire to abate, lest, no longer able to support itself, it might fall upon the furnace and be consumed, he was fain to remain in the car until half-past four in the following morning, when some workmen happening to pass by, came to his assistance, and enabled him to quit his irksome situation.

Sowden, Captain. London, 28th June, 1802.

Sparrow, Mr. Oxford, 13th June, 1823.

Spencer, Mr. E. London, 23rd May, 1836.

Spinney, Mr. Cheltenham, 3rd July, 1837.

Spinney, Mr. Junior. Gloucester, 10th November, 1836.

Spooner, Miss. Bolton, 30th August, 1826.

St. Albin, Mr. London, 1st July, 1824.

Steel, Mr. Thomas. Warwick, 8th September, 1836.

Stephenson, Mr. W. Blackburne, 7th April, 1828.

Sternberg, Joachim, Comte de, Prague, 31st October, 1790.

Stocks, Miss. London, 25th May, 1824.

Strapps, Mr. T. W. Manchester, 3rd August, 1827.

Stuver, Herr Gaspard. Vienna, 25th July, 1784.

T

TALBOT, The Honorable W.   London, 6th October, 1836.

TALBOT, Baroness.   London, 6th October, 1836.

TAYLOR, Mr.   Manchester, August, 1832.

TAYLOR, Mr.   London, 26th June, 1837.

TESTU-BRISSY, M.   Paris, 18th June, 1786.

THIBLE, Madame, the first female aeronaut, ascended from Lyons in a *montgolfiére*, 4th June, 1784, in company with M. Fleurand, in the presence of the court and of the King, Gustavus of Sweden, then travelling under the fictitious name of Count Haga.   Madame Thible is perhaps the only woman who ever ascended in a fire-balloon.

THOMAS, Mr.   London, 4th September, 1837.

THOMPSON, Miss.   London, 29th June, 1814.

TOLLEMACHE, Captain.   London, 23rd June, 1837.

TRACEY, Mr. H.   London, 16th June, 1837.

TRAVIS, Mr.   Manchester, 18th July, 1832.

TRUCHON, M.   Javelle, 25th August, 1785.

TUMERMANS, Mademoiselle Von.   Metz, 2nd June, 1788.

TURNER, Mr. F.   Cambridge, 19th May, 1832.

TYTLER, Mr. James, made an ascent in a *montgolfiére*, or fire-balloon, from Comely Gardens, Edinburgh, on the 27th of August, 1784, in virtue of which he is entitled to the triple distinction of being the *first* native of Great Britain that ever navigated the skies; of having accomplished the *first* aerial voyage ever executed in these realms, and (with the exception of Mr. Sneath's,) the *only* one upon the principle of the original inventor, in which the agent of the ascension was atmospheric air rarefied by the application of artificial heat.   As this event has either been disallowed or over-looked by all those who have hitherto professed to chronicle the progress of aerostation, we shall take the liberty of quoting the words of an eye witness by whom it is thus described in a letter dated the day of the occurrence, and inserted in the " London Chronicle," the following week :—

" Edinburgh, August 27, 1784.

" Mr. Tytler has made several improvements upon his fire-balloon. The reason of its failure formerly was its being made of porous linen, through which the air made its escape. To remedy this defect Mr. T. has got it covered with a varnish to retain the inflammable* air after the balloon is filled.

" Early this morning this bold adventurer took his first aerial flight. The balloon being filled at Comely Garden he seated himself in the basket, and the ropes being cut he ascended very high, and descended quite gradually on the road to Restalrig, about half a mile from the place where he rose, to the great satisfaction of those spectators who were present. Mr. Tytler went up without the furnace this morning; when that is added, he will be able to feed the balloon with inflammable air, and continue his aerial excursions as long as he chooses.

" Mr. Tytler is now in high spirits, and in his turn laughs at those infidels who ridiculed his scheme as visionary and impracticable. Mr. Tytler is the first person in Great Britain who has navigated the air."

Notwithstanding the unquestionable testimony which is afforded in this, as in several other periodical documents of the day, it is singular to observe with what perverseness the various writers upon the subject have contrived to avoid the admission of this ascent, and concur in ascribing to a foreigner the merit of having accomplished the first aerial expedition ever executed in this country. Nay, such is the extent to which this prepossession has operated, that in one publication in

---

* The application of the term " inflammable" here, is evidently a mistake of the writer, arising from an ignorance of the real meaning of the word, and an incorrect association between the material and the cause of its production.

which both adventures are accurately and chronologically detailed, the honour of the priority is assigned to Signor Lunardi, by the *letter* of the text, which is confirmed to Mr. Tytler, by the *dates* under which they are respectively recorded.[*]

It is true the ascent in question was not attended with any of those remarkable circumstances by which the exploits of the earlier aeronauts were generally signalized; neither was the distance run over, nor the rate at which it was accomplished such as to entitle it to particular notice on the score of these attributes. To regulate the merits of an ascent according to such a scale, would, however, be most unjust; these are, in fact, matters depending entirely upon circumstances over which the individual has no control; and many instances might be quoted of experiments remarkable enough in other particulars, which, in these, might be considered as singularly deficient. Were such, in fact, to be taken as the test of admission to the honours of aerostation, de Rosier and Arlandes must relinquish the glory of the first aerial flight, whose utmost stretch was only 5,000 toises; nay, even the celebrated ascent of Montgolfier himself, at Lyons, must be erased from the list, in which the distance accomplished was even yet more inconsiderable.

V

VALE, The Reverend B.   Hanley, 3rd October, 1826.

VALLET, M.   Javelle, 25th August, 1785.

VARIN, M.   Rennes, 15th August, 1801.

VERNON, Admiral Sir Edward.   London, 23rd March, 1785

VEYSEY, Mr. J.   Manchester, 4th November, 1837.

VIPOND, Mr.   Sunderland, 1835.

VIRLY, M. de.   Dijon, 12th June, 1784.

VIVIAN, Mr. E.   London, 29th August, 1835.

VOIGT, Herr.   Huddersfield, 13th October, 1828.

[*] See Gentleman's Magazine, Vol. LIV. Part II. pages 709 and 711.

W

WARBURTON, Mr. E.   Cheltenham, 22nd September, 1836.

WARWICK, Mr.   London, 18th May, 1837.

WATSON, Mr.   London, 12th October, 1837.

WATTS, Mr.   London, 26th June, 1837.

WEBB, Mr.   Bath, October, 1823.

WEBB, Mr.   London, 9th August, 1837.

WEDGEWOOD, Mr.   Newcastle, 26th October, 1826.

WESTCOTT, Mr. J. H.   London, 2nd July, 1833.

WESTCOTT, Mr. P. T.   London, 10th June, 1829.

WHITCHER, Mr.   Southampton, 6th August, 1829.

WHITE, Captain.   London, 25th July, 1836.

WHITTAKER, Mr.   London, 21st July, 1826.

WILCOX, Mr. James.   Philadelphia, 28th December, 1783. The first aeronautical experiments instituted in America, or indeed any where out of the country of their original invention, were those of Messrs. Rittenhouse and Hopkins, members of the Philosophical Academy of Philadelphia, immediately after the first publication of the exploit of de Rosier and Arlandes; and, indeed, it may almost be said coetaneously with those by which the discovery itself, in the elder quarter of the world, was signalized and confirmed.   The singularity of the plan upon which the experiments in question were contrived, no less than their priority, entitles them to notice.   Instead of the one large vessel to contain the buoyant principle of the ascension, the machine employed by these gentlemen consisted of several small balloons, in all amounting to no less than forty-seven, connected together and attached to the car, or *cage* as they termed it, intended for the accommodation of those by whom it was to be freighted.   After several preliminary experiments in which animals, and, in one case, a man was let up to a certain height and drawn down again by means of a rope, Mr. James Wilcox, a carpenter, who had been induced for a small sum of money to hazard the attempt, entered the car, and the ropes being severed, rose into the air, where he remained nearly ten minutes, when perceiving himself rapidly approaching the Schuylkill river (which is there of a considerable breadth), and apprehensive of falling into it, he took the necessary steps to occasion his descent. For this purpose, according to his instructions, he made incisions in three of the balloons which were nearest to him with a knife, which not proving sufficient he immediately opened three others. Seeing, however, that the machine did not descend, and fearing he should not be able to clear the river, he hastily and incautiously opened five more all together, and upon the one side, by which means the equilibrium of the machine was destroyed, and he reached the ground with so much violence that his wrist was dislocated in the fall.

If we regard the date of this experiment and that of Messrs. Charles and Robert upon the same principle, and take into account the distance between the two localities, we shall be led to the adoption of a conclusion most creditable to the scientific skill of our transatlantic brethren; namely, that the application of hydrogen gas to the purposes of aerostation was originally demonstrated in America without any communication with the continent of Europe; and that consequently had it not been for the restrictions which the interval between them necessarily imposed upon their intercourse, the old world might have had to thank the new for the establishment of the art upon the principles upon which it is now universally conducted.

WILLIAMS, Mr. W. H.   London, 8th May, 1827.

WILLERTON, Mr. J.   Boston, 7th May, 1828.

WINDHAM, Rt. Hon. W.   Moulsey-Hurst, 5th May, 1785.

WOOD, M. P.   Wakefield, 9th September, 1828.

WOODHOUSE, Dr. J. T.   Cambridge, 16th May, 1831.

WOODS, Mr. J.   Stroud, 12th October, 1836.

WOODROFFE, Mr.   London, 21st September, 1836.

WROTTESLEY, Mr. J.   London, 9th August, 1836.

Y

YARMOUTH, Lord.   Paris, 19th December, 1836.

YOUNG, Mr.   London, 21st September, 1836.

Z

ZACHAROF, M., a Russian gentleman, ascended in company with a French professional aeronaut, of the name of Robertson, under the immediate direction of the Imperial Academy of Sciences at Petersburgh, June 30th, 1804. The express object of the ascent, as might indeed be conjectured from the nature of the patronage under which it was executed, was the institution of experiments and observations in different branches of science. Accordingly of such a nature is the reputation it has obtained among those who are not practically acquainted with the art, the details of which it necessarily involves. To those, however, who examine it under such advantages, a variety of circumstances suggest themselves which incontestibly prove, either that the results of the expedition have been most grossly falsified, or else that the parties themselves must have been deceived in their speculations to a degree never before witnessed in the annals of human credulity. We shall take the liberty of investigating a few of the positions they have endeavoured to establish; more indeed because the examination of them will serve to give a further insight into some of the peculiarities of the art we have taken upon us to illustrate, than for any benefit it is calculated to confer upon science in general.

Among other circumstances upon which they dilate, and from which they declare themselves to have obtained most favorable results, were two mechanical contrivances designed with a view to mark the progress of the balloon; the one partaking of the nature of the nautical log, the other a telescope, suspended vertically through an aperture in the bottom of the car. The former of these consisted of a light cross formed of two sheets of tissue paper, fixed at right angles by means of a frame of light wood, and let to hang over the side of the car, at the end of a line 60 feet in length. The mode in which this machine is said to have operated is by the difference presumed to exist in the motive energies of two bodies of different bulks moving at liberty in the same medium. It is scarcely necessary to observe, that the imputed result is entirely a fabrication. No such difference exists or can be made to exist, between bodies, no matter how constituted or disposed, freely suspended in the same current. Nor can the assertion be palliated by ascribing it to an error on the score of a difference of current; the distance (60 feet), is much too small to admit of the bodies being at the same time in such circumstances; while even if they were so, the results would have been so varied as must at once have led to the detection of the cause, had the parties been ever so little acquainted with the principles of the science in the service of which they were enlisted.

With regard to the second of the above contrivances, the object of which was to ascertain the rate of the balloon's progress, by a trigonometrical observation of the portion of the earth's surface passed over in a given time, the fallacy of their assertions, though not perhaps so generally apparent, is even still more heinous and inexcusable. All those who have ever attempted to look through a telescope from the car of a balloon will bear me witness to the utter impossibility of ever obtaining the slightest information through the intervention of that instrument. The motions of the balloon are not only so irregular, but being totally unfelt, so impossible to provide against by any disposition or adaptation of the body, (in the manner it is effected at sea), that not even with the greatest exertion of care and skill can any thing be observed for a second even, in the act of passing across the field of view, much less submitted to a studied or continued investigation, with the accuracy necessary to admit of its being adopted as a ground for the deduction of any inference of the kind in question. How they came to advance assertions so utterly fallacious can only be accounted for upon the supposition, that satisfied of the correctness of their conclusions in a theoretic point of view, and not apprehending any impediment in their practical illustration, they never troubled themselves with submitting them to a trial, but taking it for granted the results would be what they anticipated, boldly gave out as facts what were only so in their own mistaken estimations.

The alleged inability of the birds they let loose to support themselves at an elevation of about 8,000 feet, and their consequent precipitation to the earth, is another error, which the united testimony of all who can be relied upon decidedly refutes, and which nothing indeed but a vain-glorious desire of temporary praise, backed by an utter disregard to truth and recklessness of the consequences, could ever have induced them to admit.* Other circumstances might likewise be noticed, equally at variance with probability and fact, but that enough has, we believe, been already adduced to prove how groundless are the conclusions to which they pretend to have arrived, and how little deserving of any share of confidence or consideration which may have hitherto unjustly been bestowed upon them.

ZAMBECCARI, Il Conte, one of the most deservedly celebrated of those whose names adorn the annals of aerostation, was a Bolognese nobleman, the friend and companion of the celebrated Kotzebue, and equally remarkable for the novelty and ingenuity of his views as for the boldness and zeal with which he endeavoured to realise them. In the very outset of the art, of which he afterwards became the victim, we find his name coupled with the ascent of a small balloon, 10 feet in diameter, which he made and launched in London, on the 25th of November, 1783, generally related as the first aeronautical experiment ever instituted in this country.† Here likewise he made his first ascent, on the 23rd March, 1785, in company with Admiral Sir Edward Vernon; upon which occasion, after having proceeded to a considerable altitude they descended at Horsham in Sussex, a distance of 35 miles in something less than an hour.

* Upon this head, consult the observations contained in Appendix F.

† The precedence ascribed to this event can, however, have been but of very few hours at the utmost; on the same day another balloon, which had been manufactured by an ingenious Prussian, M. Argeue, was launched from the hands of the king himself at Windsor, for the entertainment of his royal consort. See *Scots' Magazine*, date as above.

Shortly after this he returned to his native country, where he devoted himself to the study and practice of aerial navigation, with a zeal and ingenuity which never abated until they were finally quenched with his life in the last fatal enterprise which terminated his career. From such a man, with such a disposition, results of an ordinary description were not to be expected. Endowed with a courage, amounting in fact to heroism, and ever aiming at objects, unattainable but by the exercise of the highest qualities backed by the most favourable disposition of circumstances, almost all his exploits concluded with events which would have quenched the ardour and determined the efforts of any but himself. Twice successively, in the years 1803 and 1804, did he descend into the Adriatic, where he remained from five to seven hours; the first time reduced to the last extremity from the rigorous inclemency of the season in which he had thought proper to institute his experiment; the second time, almost scorched to death by the accidental combustion of his balloon, which ensued upon a violent dislodgement of the fire in an unsuccessful effort to terminate his excursion.

Similar in every respect to that above-mentioned, save in its issue, was the accident by which a few years later his death was so unhappily occasioned. Attempting to descend after a voyage he made from Bologna, on the 21st September, 1812, in a balloon upon the same pernicious principle as the former, accompanied by a certain Signor Bonaga, the grapnel caught for an instant in the branches of a tree, and by the shock which it produced occasioned the subversion of the fiery contents of the furnace, when the whole machine became speedily enveloped in flames. Driven to desperation by the rapid encroachment of the devouring element, after all their efforts had proved unavailing to arrest its progress, as a last resource they sprung from the car, regardless of the elevation they were now rapidly regaining. Zambeccari was killed upon the spot: his companion more fortunate, though fearfully injured, escaped with his life.

Zichy, Il Conte. Paris, 9th January, 1837.

₊ The following names (with the exception of the first, which was accidentally overlooked in the commencement) were excluded from their proper places owing to the sheet to which they belong (marked S in the preceding catalogue) having been worked off by mistake before it was completed.

BLAKE, Captain. London, 3rd June, 1785.

GENISTE, M. Paris, 9th January, 1837.
GLOSSOP, Mr. W. Sheffield, 4th September, 1828.
GREEN, Surgeon. London, 15th September, 1837.
GREEN, Mr. Manchester, 4th November, 1837.
GREGORY, Mr. Oxford, 8th June, 1837.
GYE, Mr. Senior. London, 14th August, 1837.

HARMAN, Mr. Uxbridge, 6th October, 1835.
HINES, Miss. Beccles, 11th October, 1785.
HITCH, Mr. S. Gloucester, 15th October, 1836.
HOLT, Captain R. Wigan, August, 1828.
HORTON, Captain Wilmot. London, 23rd June, 1837.

JEARRAD, Mr. R. Cheltenham, 3rd July, 1837.
JEPHSON, Mr. London, 15th May, 1837.
JILLARD, Mr. Bristol, 25th July, 1825.
JOLLIFFE, Mr. Sevenoaks, 23rd August, 1825.
JULLIEN, M. Paris, 9th January, 1837.

KRASKEWIL, Herr. Vienna, 4th May, 1812.

LAING, Mr. London, 16th June, 1837.
LAMBERT, Mr. London, 30th May, 1837.
LAWSON, Mr. Keighley, September, 1829.
LEBERRIER, M. Paris, 28th July, 1832.
LENNOX, M. Le Comte. Paris, 28th July, 1832.

LENNOX, Madame.   Paris, 28th July, 1832.

LORD, Mr. J.   Manchester, 4th November, 1837.

MANSFIELD, Mr.   Manchester, 4th November, 1837.

MENNER, Herr.   Vienna, 4th May, 1812.

Moss, Mr. J.   Cheltenham, 3rd July, 1837.

MUSGRAVE, Mr. W.   Leeds, 19th October, 1837.

NEWMARCH, Mr.   Halifax, 9th August, 1785.

PARKINSON, Mr.   Bury, 27th June, 1828.

PAUMIER, Mr.   Whitehaven, 3rd September, 1832.

PEARNE, Mr. G.   Dover, 10th September, 1828,

PEILE, Mr. W.   Whitehaven, 3rd September, 1832.

PEMBERTON, Mr.   Preston, 14th June, 1821.

PENNY, Mr.   London, 4th July, 1825.

POOLE, Mr.   Bury, 15th October, 1785.

POOLE, Mr.   Preston, 21st June, 1821.

PUGH, Mr. W.   Gloucester, 15th October, 1836.

RAPHINE, M.   Brentford, 22nd November, 1784.

REDMAN, Mr.   1st June, 1786.

REID, Mr.   Perth, 1st July, 1831.

_____

Besides the foregoing, fourteen persons have ascended, whose names do not appear in the only documents in which the circumstances of their several exploits have been preserved. Of these, three ascended with Herr Stuver from Vienna, 25th July 1784,—one, already alluded to as having accompanied the Duke de Chartres, in the ascent from St. Cloud, on the 15th of July, 1784,—one, a servant of Signor Lunardi, who took his place in the balloon when, through an accident, he was incapacitated from ascending himself at Chester, about the 25th of August, 1785,—one with M. Adorn, from Strasbourg,

15th of May, 1784,—one, a young lad, drummer in the garrison of the town, with M. Rousseau, from Navan, 14th of April, 1784,—one, also a young boy, who happened to be seated in the car of Mr. Harper's balloon, when it accidentally slipped from its moorings, 31st of July, 1785,—two in company with Mr. Binn, from Halifax, 9th of August, 1785,—one with Mr. Sadler, from Birmingham, 7th of October, 1811,— and three, one a Persian physician, and two Bostangis, or officers belonging to the seraglio, at Constantinople, early in the year 1786. The circumstances of this experiment are very interesting, considering the age in which it took place and the country in which it was accomplished. They ascended from the court attached to the palace of the Sultan, in the presence and to the great delight of the ladies of his harem, and traversing the sea which divides the European from the Asiatic continent, after an agreeable and prosperous voyage of four hours and a half, descended about thirty leagues from the coast, in the middle of the castle of Bursia, the residence of a pacha, by whom they were nobly and hospitably entertained. Upon their return they were received with all the honours which an admiring public could bestow, and the balloon itself, as a memorial of the exploit, was ordered by the Sultan to be suspended in the Church of St. Sophia, where I believe it remains to the present day. This is the only instance upon record in which the passage from one quarter of the globe to another has ever been made by means of the balloon.

These are all that I am at present aware of that have ever ascended into the bosom of the atmosphere detached from the surface of the earth, or as the French not inaptly term it, *à ballon libre;* in contradistinction to those occasions wherein the machine has been retained by ropes, and which with equal felicity have been technically denominated, *à ballon captif.*

Beyond the omissions occasioned by the premature working off of one of the sheets, before alluded to, (and which will be found supplied in the general list of corrections at the commencement

U

of the work), some deficiencies will no doubt have been observed in the specification of the dates and places of several of the ascents, which, for the satisfaction of those who might be inclined to enquire how it were possible to have obtained certain assurance of such exploits independant of the particulars in question, it may be advisable to explain. The fact is, that many of the names referred to having been obtained from newspaper-cuttings or other documents of a similar description in the cabinets of the curious in these matters, in which, from the neglect of those by whom they were originally extracted, the date and place of the experiment have not been preserved, all clue to those particulars was either entirely lost or only to be roughly computed by a reference to those of some other incident better known in these respects, and of which some trace on the back or otherwise accidentally happened to be retained. Perfect accuracy is, in truth, scarcely compatible with a registry of events of more interest than importance, extending throughout so large a portion of time, and embracing so many different countries, some of them unprovided, (until lately, at least,) with any of those periodical sources of public information, by means of which alone the remembrance of such circumstances are properly or correctly preserved. Should the name of any gentleman, however, happen to be accidentally omitted, or any deficiency appear in the specification of the place or date under which it is recorded, if he will only take the trouble to advise me of it through the medium of my publisher, he may rely upon its being corrected in a future edition; that is to say, should the favour of the public extend so far as to afford me the opportunity required.

APPENDIX D.

OBSERVATIONS UPON THE MECHANICAL DIRECTION
OF THE BALLOON.

To display in its proper colours the long-contested
question of aerial navigation, and enable the general
reader to form an opinion for himself as to the probability
or improbability of the accomplishment of that most interesting, and indeed important of all mechanical desiderata, is
the object we have proposed to ourselves in the following
investigation. In the execution of this design we have
felt it necessary to abandon the attractive but irregular
paths of description for the more tame and tedious avenues
of systematic reasoning. But the truth is, the enquiry
itself properly admits of no other mode of treatment.
The case of a balloon artificially propelled through the
air, is one essentially involving the elements of the pneu-
matical and mechanical sciences, and can only be satisfactorily argued as to its practicability, upon the basis of
strict mathematical induction. Any attempt to dispose of
it without these aids, however it may serve to amuse,
must notoriously fail in the only object for which its
services are required; namely, to determine the expectations of the curious, and direct the efforts of such

among them as may yet feel inclined to indulge in the attempt to accomplish it.

The recondite nature of the principles upon which it is based does not, however, by any means involve the necessity for a like abstruseness in the conduct of the enquiry they are designed to support; nor, indeed, would such a conclusion have accorded with the purposes we have in view. They are not the learned but the unlearned that our labours are intended to enlighten. To those who are themselves versed in the sciences that bear upon the case, the following observations (with the exception of a few remarks which a practical acquaintance with the art has specially enabled us to supply) can possibly present nothing new; nothing, in short, with which they are not better acquainted, and of which they are not better able to judge than ourselves. It is to the general reader alone that we address ourselves, who with equal capability of drawing conclusions, may haply be devoid of a proper knowledge of the grounds whereupon to construct them.

With this view we have studiously endeavoured to avoid the employment of all such terms of art as are not in use in common parlance, and otherwise to adopt a style and method as familiar and concise as is consistent with the clear exposition of the subject we have taken upon us to illustrate. In accordance with these principles, our intentions in the following investigation are to ascertain and define,—first, the obstacles which interfere with the active progress of the balloon;—secondly, the mechanical means required to surmount them;—thirdly, the natural power by which those means are to be put in operation;—and, lastly, to point out certain regulations and restrictions by which they must be governed in their application in order to be really available to the purposes for which they are designed.

By this method of proceeding, one important conclusion at any rate we shall have established; namely, what are the means by which *alone* the direction of the balloon can ever be accomplished. Under what particular form these means may be applied, or whether indeed their application is within the reach of those powers which Providence has placed at our disposal, we leave entirely to the judgment and ingenuity of the reader himself to determine.

(I.)

The moment a balloon has cast off its last hold upon the solid earth and been received into the bosom of the air, it becomes at once, and, in the absence of all foreign interference, completely subservient to the same impulses and affected by the same impressions as those which govern the disposition of that element itself. To the actual amount of these, the varied and inordinate rate of the atmospheric currents, is to be attributed the entire of the difficulty that involves the question of aerostatic guidance. The mere tenuity of the medium, the want of a consistency sufficient to afford grounds for the establishment of a proper *point d' appui*, or fulcrum for the application of the requisite forces (which by most persons is inconsiderately regarded as the great obstacle to success), however it may avail to enhance the difficulties of pure mechanical flight, is literally of no importance whatever as regards the artificial propulsion of the balloon. The cases in this respect are entirely dissimilar. In the one, a force (the attraction of gravitation) has to be overcome by another, (the resistance of the atmosphere) with which it has no connexion, and which, therefore, there is no reason to suppose necessarily competent to the charge : in the other, the forces to be overcome and the means of overcoming them are the same,—namely, the resistance of the atmosphere; in proportion as the grounds of propulsion are feeble, the opposition against which they have to contend, and by which they are regulated in their amount, are feeble also.

Were it not, therefore, for the rate of the medium and the obligations it imposes upon the conduct of the operation, nothing would be simpler nor more certain than the mechanical direction of the balloon. Action and reaction being invariably equal, any exertion of the proper means, no matter how slight, must inevitably produce a determinate advance in its position ; and that, without any regard to the direction of the medium in which it is conveyed.* It is true that where the dispro-

---

* In considering the case of a body advancing through the air, under the exercise of means of propulsion inherent in itself, the reader will bear in mind that neither the rate nor direction of the medium in which it is conveyed in any way affects its condition, or occasions it to suffer any sort of violence beyond what, with the same exertions on its part, it would experience were it to seek to advance *with* instead of *against* it. The idea of a vast and cumbrous machine struggling to maintain itself in the teeth of a rude and impetuous current, is likely to be a very different one from that of the same body calmly exercising the *same* force with the advantage of the wind to second its exertions. The distinction, however, so far as the condition of the body is concerned, is a false one. Differing in this respect from other locomotive machines, all the forces by which it is operated upon are determined by its own exertions alone, proportioned to the rate and opposed to the direction which they seek to establish.

portion between the resisting powers of the means of propulsion and those of the machine whose movement was to be the result of their operation was extremely great, its actual progress would be extremely small; some, however, little as it might be, would positively be realised, and the only question would be how far the advantages obtained were worth the exertions employed to secure them.

With an independant motion, however, in the medium of its conveyance, the guidance of the balloon to any extent is by no means a necessary consequence of any exertion of forces with which it might be possible to invest it; and this it is which constitutes the great difficulty by which aerial navigation is beset, and by which it is so unfavourably distinguished from almost all other known modes of transport. If a steam-engine, for instance, should be competent to propel a carriage even at the slight rate of only a mile an hour, still the means employed might be considered as successful to that extent at least, and the machine, though comparatively inefficient, yet, as far as it went, available to purposes suited to its power. Such, however, is by no means the case with the balloon; the progress conferred upon it by foreign forces, be it ever so great, can never be set down as so much gained, nor can the means of its propulsion be considered as successful to *any* extent that are not so to a *given* one. Acting in and under the influence of a medium, itself endowed with rapid motion, a very considerable degree of velocity might be acquired by the balloon without any actual gain at all; and were we to take extreme cases, the greatest rate of motion ever enjoyed by any terrestrial object might be conferred upon it, and yet so far from advancing it might be absolutely a loser in point of space from where it was ere it commenced its career. Before the balloon, therefore, can make sure of obtaining any advantage whatever from the exercise of its means of propulsion, it must be able at all events to command a rate of motion superior to that of the medium in which it is conveyed.

The movements of the atmosphere, with which alone we have here any concern, are, as we all know, a most variable quantity, comprising within their limits almost every degree of velocity with which we have any practical acquaintance, and pervading (so far as we have any right to conclude) all those regions which, from their proximity to the earth, constitute the proper sphere of the balloon.

I am aware that an opinion is very prevalent among aeronauts, and which is also favoured by some meteorologists of distinction, (especially those of Germany and France), that all these changes are confined to the lower regions of the atmosphere, and that beyond a certain elevation, a state of perfect, or at least comparative, tranquillity may be looked upon as the natural condition of the etherial space. To what to ascribe the origin of this opinion I am totally at a loss to conceive, unless indeed it may be to that innate disposition in men to believe what they desire to be true, and to adopt, without questioning, whatever appears to favour their particular predilection. The supposition, however, is by no means borne out by facts; on the contrary, many instances might be adduced from the registered annals of the art, in which considerable excitement has been found to prevail in the upper regions of the atmosphere, at a time, too, when, comparatively speaking, no motion whatever could be perceived in the portions adjacent to the surface of the earth. In one of the two ascents which Signor Lunardi executed from Heriot's Hospital Grounds, at Edinburgh, notwithstanding a state of perfect tranquillity uniformly prevailed below, the rate of the balloon's course at the greatest altitude to which he arrived exceeded 70 miles an hour. On the 28th of April, 1802, Captain Sowden, in company with M. Garnerin, ascended from the Ranelagh Gardens, near London, and after continuing at a very considerable elevation, in three quarters of an hour descended near Colchester, a distance of 60 miles; having thus accomplished a rate of motion equal to 80 miles an hour, although scarcely any could be perceived at the immediate surface of the earth. A still more striking proof of the existence of rapid atmospheric currents at excessive elevations, and one which appears to be decisive on the subject, is afforded in the second ascent of M. Gay-Lussac from Paris, in which a very considerable rate of motion was accomplished, although the whole of the voyage, with the necessary exception of the ascent and descent, was conducted at an altitude bordering upon 23,000 feet, the greatest to which any balloon has hitherto been known to arrive.* It is unnecessary to multiply examples to disprove the truth of a general rule ; enough has already been adduced to determine the fact, that at the greatest elevation ever attained by man, very considerable atmospheric currents have been proved to exist. What may be the case at a still higher elevation we must leave to future experience to determine ; in the mean time, we must continue to regard the atmosphere as we have found it ; and in our treatment of the question before us, consider the aerial vehicle as liable to the influence of those forces which have hitherto proved superior to all the efforts by which it has been attempted to subdue them.

These forces then are, as I have said before, of a very variable disposition, embracing within their limits almost every degree of motion with which we are practically acquainted, from a state of perfect quiescence to the enormous rate of one hundred miles an hour. Such a rate of motion, it is true, is very uncommon, and, in our climate at least, of such rare occurrence that it could not be imputed as a valid objection to any plan for the guidance of the balloon, that it was not calculated to meet so extreme a case as that which we have here specified. The average rate of the wind in these climates (which we have chiefly in view in the following treatise), may be said to be about 25 miles an hour ; this we are enabled to determine, not from the observations of the meteorologist alone, but, (what is more to the point, because founded upon experience in a part of the atmosphere with which we have more especially to do,) from a consideration

* See Appendix C, article " Gay-Lussac."

of the average rate of Mr. Green's aerial excursions, deduced from a series of 249 voyages, executed generally in the most favourable periods of the year. From this we learn that 25 miles an hour* is the mean rate at which a body floating in the atmosphere may be expected to be transported; and with resources to that extent would it be necessary to be provided, were the *average* amount of the obstacles to be taken as the measure of the means to be employed in surmounting them.

But the average amount of the antagonist forces, however it might be deemed a sufficient guage in the case of other locomotive machinery, could by no means either prudently or properly be admitted as an adequate allowance in that of the aerial conveyance. The powers by which the progress of the balloon is liable to be affected are so vast, that were she only provided with the means of resistance upon so limited a scale, the deficiency in extreme cases would involve consequences far beyond what the exercise of her own resources could ever enable her to retrieve. No arugment can be drawn from a consideration of what would be reckoned sufficient in other cases (in marine navigation for instance), to sanction the admission of the same scale whereon to measure the means required for the guidance of the balloon. The extreme rate of a current at sea, never, I believe, reaches 10 miles an hour :† that of the atmosphere in motion, I have before observed, occasionally amounts to 100 miles in the same time. The actual consequences, therefore, to a ship furnished

* The total distance which Mr. Green accomplished in the course of his first 200 aerial excursions, a very accurate computation enables him to fix at 6000 miles; and the time consumed in the performance at 240 hours. The former of these two quantities divided by the latter gives the quotient above mentioned.

† The currents proceeding from the action of the tides, which occasionally accomplish a much higher rate of motion, are not, nor should they be, here taken into account; inasmuch as, from their very nature, alternating successively in two opposite directions, they invariably neutralise their own influence every twelve hours, and cannot really be said to have any effect upon the course of a vessel whose voyage is intended to endure for more than half the above period.

with means equivalent to half of what she might have to
encounter, would be but trifling compared to what a balloon
would suffer in a similar emergency and similarly provided to
meet it. Each, it is true, would lose but one half of her way;
but the half of her way lost to the ship would be only equal to
five miles an hour, and the result but the retardation of a few days
at the utmost in the date of her arrival at her destined port:
the loss of half her way to the balloon would amount to fifty
miles an hour, and the probable result would be that she would
have reached the antipodes ere any circumstances might have
occurred to favour the recovery of her course.

From the consequences of an inadequacy to contend with
superior forces, the balloon again has none of those shifts to
relieve her, such as oblique sailing, tacking, or even temporarily
suspending her progress, to which the mariner can resort in
similar cases, and which enable him to put up with a comparatively inferior power. If the force which opposes the balloon,
she is unable to subdue by direct opposition, she must be content at once to submit to the consequences of defeat. This is
the more necessary to be insisted upon, because I have generally
found persons resort to such arguments, in order to bolster up
a feeble scheme of aerial navigation; flattering themselves that,
although they might not be able to accomplish a progressive
motion in direct opposition to a powerful current, they would
be able to take an angle and traverse it obliquely, as a ship
tacks against a wind; or, should that fail, come to an anchor,
and thus remain neuter during the predominance of the powers
they are unable to contend with.

The expedients, however, to which they advert are totally
inadmissible, and, with regard to the former, absurd. Tacking,
as practised at sea, is an operation requiring the presence of two
independent media, and may be defined, the taking advantage of
one of them (the water) to secure a *direction* for the exercise of
the force obtained through the intervention of the other (the air);
such a resource is as inefficient with the aid of one medium only,
as the action of the male screw would be without the female, or

the lever without its fulcrum. If a balloon cannot make head against a current of air in direct opposition to its course, it only aggravates the mischief by any attempt to meet it obliquely.*

With regard to the other expedient alluded to, namely, the temporary discontinuance of the course of the balloon whenever the condition of the atmosphere should happen to exceed its powers of resistance, the idea is replete with practical impossibilities. The moment a balloon is inflated, the worse the weather the more urgent is the necessity for her immediate departure; every moment she delays, teems with risk, and should the forces in question be excessive (which, indeed, is the very contingency contemplated in our argument), the only chance of her security is in the air. These are objections which the inexperienced

* The examination of the following Diagram will render this conclusion more apparent :—

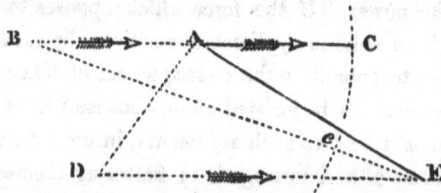

Suppose a body freely suspended in the air and capable of accomplishing a rate of motion equal to *ten* miles an hour, were to set out from the point A, with the intention of proceeding in the direction of A B, against a wind moving at the rate of *twenty* miles an hour; by the time it had attained the point B in the body of the atmosphere, that point itself would have been transferred with the progress of the medium to a spot corresponding to C upon the surface of the earth; the course of the body would be represented by the line A C, and the loss of way would be equal to the difference between the two rates. Were the body, with a view to avoid the direct opposition of the air, to take an angle and seek to advance in the direction A D, by the time it had reached the point D in the body of the medium, that point would have been transferred to E on the surface of the earth; the course of the body now would be indicated by the line A E, and the station it had acquired would be further removed by the distance *e* E from the point B, which it had first aimed at, than if it had proceeded thither in direct opposition to the wind.

reader cannot be expected to appreciate, but which all those who have any practical acquaintance with the details of the art will be ready at once to admit. If they are conclusive against the possibility of adopting the step here alluded to, with a balloon of the ordinary simple principle and advantageous construction, how much more so must they be when applied to one fitted up with the vast and cumbrous apparatus required for its propulsion, increasing the liability to damage exactly in the ratio of the inability to resist it. The expedient, in short, is one which never could be resorted to except when it was unnecessary, and never could be necessary except when it was impracticable.

The extreme rate of motion with which it may fairly and reasonably expect to have to contend, must, therefore, be had in view in all schemes which propose to render the balloon a certain and serviceable mode of transport, and at any rate as much of it provided against as shall leave a deficiency within the reach of her own resources to repair.

From a consideration of all the bearings of the case, and desirous as much as possible to favour the hopes of an aerial navigation, I am bound to say that unless the balloon can command a rate of motion equal to 30 or 35 miles an hour, it cannot claim to be considered as a mode of transport applicable to useful purposes, or on a par in point of advantages with any of those whose services it might be expected to supersede.

Now all this velocity it is evident cannot be accomplished without the development of a certain force of resistance, which is in fact the very measure of the difficulty we have hitherto been labouring to ascertain. This resistance is chiefly of two kinds; the one, the direct impact of the atmosphere,—the other, the friction occasioned by the action of its particles along the surface of the opposing body; both of which are determinable as to their amounts by a consideration of the form of the object and the rate at which it is impelled.

1. The former of these, the direct impact of the atmosphere,

is by far the more serious obstacle of the two, and that against which the efforts of the aerial engineer have hitherto been almost exclusively directed. In a previous part of this work[*] we have taken some pains to point out in what manner and to what extent the form of the body is capable of modifying this force, and have within certain limits established a rule by which to determine the comparative amounts of atmospheric resistance upon bodies opposing plane and conical surfaces to its action. To avoid, therefore, entering anew upon the same ground we shall only observe generally, that from one-half to one-third less opposition is realised by a hemisphere, or cone of equal altitude with its base, in passing through the air, than would be experienced by a plane surface equal in extent to its largest section, taken at right angles to the direction of its course. The conditions of this latter force (I mean, of the resistance afforded by the atmosphere to the impact of a *plane* surface), have already been pretty accurately investigated, and its amount, corresponding to the rate of the medium, determined by experiments ingeniously devised and carefully instituted, for all degrees of the scale, from one to one hundred miles an hour. It is scarcely necessary to observe, that whether the impact be effected by the motion of the body or simply that of the medium, the result, as far as concerns the amount of force produced, will be the same ; and that, consequently, the pressure of the atmosphere, as displayed in the phenomena of the winds, may be taken as a correct measure of the resistance which, at the same rate, the balloon would occasion for itself, were it alone to be endowed with motion. For the benefit of those who may feel inclined to enter more at large into the calculations connected with the subject, a specification of this force, as determined by the expe-

---

[*] See Letter addressed to the Editor of the Morning Herald on the subject of the Parachute, marked No. III. in the Appendix B.

riments of Messrs. Rouse and Smeaton, has been subjoined below.*

From this table it will be seen, that for every square foot of *plane* surface called into action at the rate one mile an hour, the atmosphere exerts a resistance equal to five thousandths of a pound avoirdupois; a force which is found to increase accordingly with the squares of the velocities under which it is exercised. To give some idea of what this force would be in practice, let us assume the case of a balloon of known dimensions; that, for instance belonging to the managers of Vauxhall Gardens, with which the public are no doubt by this time pretty well acquainted. This balloon is a spheroid of about 60 feet in height and 50 in breadth; in computing its powers of resistance, however, we shall not much err if we regard it as a sphere, whose diameter is equal to the mean of these two quantities. Upon this hypothesis, then, the plane of its largest section would

* TABLE,

Shewing the perpendicular force of the wind under different velocities, in pounds avoirdupois, on each square foot of plane surface, computed from experiments of Messrs. Rouse and Smeaton.

MILES PER HOUR.	PRESSURE PER FOOT.	MILES PER HOUR.	PRESSURE PER FOOT.
1	. 005	35	6 . 027
2	. 020	40	7 . 873
3	. 044	45	9 . 963
4	. 079	50	12 . 300
5	. 123	55	14 . 885
6	. 178	60	17 . 715
7	. 242	65	20 . 791
8	. 315	70	24 . 100
9	. 399	75	27 . 646
10	. 492	80	31 . 490
15	1 . 107	85	35 . 550
20	1 . 968	90	39 . 850
25	3 . 075	95	44 . 401
30	4 . 429	100	49 . 200

The terms of the scale answering to the rate of the wind at 6, 7, 8, 9, 55, 65, 70, 75, 85, 90 and 95 miles an hour, which have hitherto been omitted, are here supplied.

contain about 2,372 square feet, the resistance upon which, however, owing to its particular form would, as we have before observed, be only equivalent to that upon a plane two-thirds its dimensions, or about 1,581 square feet. Multiplying this sum by the amount in the subjoined table corresponding to any degree of velocity, we shall have at once, and with very considerable accuracy, the amount of the whole force by which its progress at that rate is affected; or, in other words, the resistance it would offer to the atmosphere or the atmosphere to it, were either to be arrayed against the other in motion at the rate in question. Thus, at the rate of thirty-five miles an hour, which we have already agreed to consider essential to the successful progress of the balloon, the opposition experienced would be $1581 \times 6.027 = 9,528$ pounds avoidupois, or upwards of four tons and a quarter.

The proportion between the force here computed, and the buoyant power of the balloon might, it is true, be considerably reduced, by the adoption of another form for the containing vessel, which should afford a less direct resistance to the impact of the atmosphere; such a modification as indeed would be necessary to render it manageable under any circumstances. Were, for instance, the contents of the sphere in question thrown into an envelope of the form of a cylinder capped at the extremity with cones, or an ellipsoid in length four times the diameter of its transverse section we should have a vessel equal in buoyancy to the former, (omitting the difference in the weights of their respective coverings), presenting an active resisting surface of only one-half the amount. Even here, however, where the arrangement of the parts is, I believe, the most favourable to the reduction of the force in question which it is possible to imagine consistent with the other exigencies of the case, the resistance to its progress at the rate required would be equal to 4,764 pounds.

2. To counterbalance in some degree the advantages which it is evident here accrue from the adoption of a form less favour-

able to the direct impact of the atmosphere, another force remains to be considered in the *friction* which is engendered between the surface of the body and the particles of the medium in which it moves. The introduction of this force is, in fact, the necessary consequence of the arrangements by which the other is sought to be avoided. Friction being the resistance exerted by the passage of particles *over* and *along* a given surface, in contradistinction to that occasioned by their impulse *against* it, must evidently increase in proportion as the facilities for the latter operation become lessened by the particular construction of the opposing surfaces. If an open umbrella be held point foremost towards the wind, almost the whole of the force directed against it will be that of impact; if it be now gradually closed without altering its direction, the force of impact will become converted by degrees into that of friction, and will give place to the latter almost entirely when the collapsion has become complete. In bodies, therefore, where the force of impact is paramount, that of friction is at a mininum, and *vice versâ;* in the intervening stages, alone, it is that both act in conjunction. The former of these being the more serious obstacle of the two, it evidently becomes the interest of the aerial navigator to construct his vessel in such a manner as shall leave him as much as possible the latter alone to contend with; hence, the more perfect the scheme for the propulsion of the balloon, the greater will be the share of the opposition to its progress, arising from the source in question.

To the reader not versed in the physical sciences it may perhaps seem strange that a resistance of such importance as to merit the consideration we have bestowed upon it, should be capable of being generated by the attrition of the particles of an elastic fluid of such slight consistency as that of the medium of the aerial conveyance. The influence, however, of atmospheric friction on the motion of bodies adapted to experience it, is unquestionably very considerable, and often productive of effects in cases where it escapes notice. It is this that, in conjunction with the attraction of gravitation, determines the course of the arrow shot from the bow, or the spear launched from the hand, which, from the minute and almost insensible impact they are calculated to experience, would otherwise be almost without a limit. To what extent it would operate in retarding the progress of the balloon—how much would have to be added upon that score to the opposition arising from the direct impact of the atmosphere—can only be ascertained by actual experiment in each particular instance. The purpose of our present observations is merely to point out the existence of such a force and explain in what manner it affects the question of aerial navigation by the limitation it imposes upon the unconditional reduction of the obstacles upon which its chief difficulty depends.

(II.)

To enable the balloon to maintain its course in the teeth of the opposition we have just been endeavouring to compute—or, more properly, to command a rate of motion whereof the opposition in question is the index as well as the result—it is absolutely necessary that it should be provided with the means of creating a reaction in the surrounding atmosphere equivalent to the resistance it will have to encounter in its employment; without a reaction there can be no propulsion; and unless the forces developed in the proceeding be commensurate with those by which the balloon itself is liable to be affected at the rate required, they will not avail to establish a progressive motion independent of that of the medium in which they are exercised.

There are, I am aware, other means of investing matter with motion besides that which proceeds from a reaction in the medium of conveyance; namely, by a resistance generated inherently and determined in a given direction by the particular construction of the parts. Such, in fact, is the force by which the ascent of the sky-rocket, the transmission of the cannon ball, the operation of the piston and cylinder, and lastly, the impetuous discharge of gas or steam, however different may be the *natural power* by which they are first called into action, are all accomplished. The incompatibility, however, of the principles upon which they all act, with the other essential conditions of the case, (as will be seen farther on)* will I think be sufficient to excuse the rejection of all such means from our consideration, and sanction the limitation within which we have confined the prospects of aerial navigation to the simple establishment of a reaction in the medium in which it is to be conducted.

We have already seen† how completely all atmospheric resistance is absent from the natural career of the balloon; how entirely the motions of the two bodies coincide when unimpeded by the interference of any foreign agents. From this it will be evident that no simple disposition of the parts, in the form of sails or otherwise, however effective they may be in marine navigation, can be of the slightest avail in the present question. With nothing to act upon them different from what acts upon all the rest of the body besides, they can be considered in no other light than as merely so many variations in the form of the aerial machine, and can be of no more service in determining its progress than oars, simply resting in the water without exercise, would be in affecting the course of a boat, as it drifted upon the bosom of the waves.

The reaction in question can therefore, it is evident, proceed from no passive arrangement of the parts, but must be actively engendered by the exercise of a force distinct from any to which the simple progress of the body itself is capable of giving rise. To this effect a certain extent of surface must be arrayed in motion so as to impinge upon the adjacent strata of the surrounding medium, and in the opposition it encounters, establish, as it were, a fulcrum for the leverage by which the machine is to be substantially propelled. Under whatever form the inge-

* See page 324.
† See page 128 and the following.

nuity or caprice of man may induce him to order his devices—
whether in the guise of oars, wings, or paddles, designed to
operate by reiterated percussion, rotation, or by continuous
impression, as exemplified by the involutions of the spiral sail
or vanes formed after the fashion of the screw—all resolve
themselves into this one principle; namely, the impact at a
certain rate, of a certain extent of surface against a resisting
medium.

In the construction of this force, therefore, two elementary
principles present themselves for consideration; namely, the
surface and the *motion* conferred upon it. To a certain extent
these two principles are vicarious of each other, and may
be indifferently employed to produce the same results. For
instance, if a machine, by the operation of a given amount of
surface, at a given rate, is able to communicate a certain speed,
the amount of this speed will be equally affected by an alteration in the dimensions of the surfaces in question, or in the rate
at which they are made to operate. To a certain extent alone,
however, this is true; for independently of the necessity there
is for a certain amount of each, no multiplication of the size of
the surfaces could ever enable them to impel the body at a rate
of motion greater than that with which they were themselves
at the time endowed. Thus a body impelled by means of a
series of revolving planes, as in a paddle wheel, striking the
air at the rate of ten miles an hour, could by no amplification of the dimensions of the surfaces be enabled to realize a
rate of motion exceeding ten miles an hour; inasmuch as the
moment it had attained that rate of motion, all reaction between
the surfaces and the air would cease; or if continued beyond
that rate, would be transferred from the back to the front of the
impelling surface, and operate to retard instead of advancing
the body to which it was attached.

But though a limit is thus imposed upon the extent to which
an augmentation in *size* may be made to supply a deficiency in
the rate of the impelling agents, no such limitation exists in

theory to the extent to which an increase in the *rate* of impact may be made to compensate for an abridgement in their dimensions: the smallest amount of surface being capable of realizing any amount of resistance providing the rate of its impressions be accelerated in proportion to the demand. Before, therefore, we can come to any definite conclusion with respect to the size of the agents of the propulsion of the balloon, it will be necessary that we investigate and determine the conditions by which the rate of their impact is governed. Should it appear from such an investigation that this rate is not more than it may fairly be expected to accomplish, it will then be open for consideration whether a still further increase may not be made subservient to a reduction in the size of the impelling planes. If, however, on the other hand (as, indeed, I fear will be found to be the more probable conclusion of the two) the velocity required for the fulfilment of the obligations alluded to be as much or more than is capable of being attained in practice, then will the conditions of size and rate assigned upon such grounds be the lowest in conformity with which the successful guidance of the balloon can be accomplished.

1. With regard to the *motion* of the impelling surfaces, therefore, one important point is already established; namely, that whatever may be their actual condition as to size, number, or powers of resistance, the rate of their impact must not be less at all events than that required as the final result of their operation: in other words, than the rate assigned to the balloon itself. Admitting the correctness of the conclusion to which we have arrived in the preceding section respecting the rate of the balloon, this obligation alone tends at once to fix upon the agents of the propulsion a velocity of action amounting to between thirty and thirty-five miles an hour.

In proceeding, however, to accomplish this rate of motion by the exercise of the mechanical means we have before laid down as essential to the purpose, another force becomes developed, tending to impair their efficiency and imperatively calling for a further accession to the velocity with which they are required to act. This force, which is precisely analogous to the "back water," observed at sea in the case of vessels impelled by steam, proceeds from a condition induced in the atmosphere by the action of the impelling surfaces themselves, whereby the adjacent portions become determined in a continuous stream, mainly opposed to the course of the machine, and with a velocity proportioned to the scale upon which the operation has been conducted.

The explanation of this process is extremely simple, although the circumstances by which it is attended are so complicated as to baffle any attempt to calculate the precise amount of the obstruction. In the atmosphere, as in all other fluids, elastic or non-elastic, a certain uniformity of condition necessarily obtains. Whenever this uniformity happens to be disturbed, either by natural causes or the interference of foreign agents (as in the present case, by the rapid action of large resisting planes), a general tendency is immediately manifested in all the neighbouring parts to rush in and restore the equilibrium; in the course of which all the same symptoms are evolved, and the same effects produced as would attend the exposure to a natural current of air. The consequences of this disturbance upon the progress of the balloon are too apparent to need explanation, although the extent to which it would be necessary to increase the speed of the impelling surfaces, in order to counteract its influence, and enable them to realize the required momentum, would not be possible without actual experiment precisely to ascertain. From a critical consideration of the several circumstances of the case, however, I do not hesitate to conclude that an augmentation of at least thirty per cent. in the rate of the impelling agents would not be more than enough to compensate for the disadvantages under which they labour upon this account, and place the machine, as far as regards the efficiency of its means, upon a par with what it would be were no such obstruction the natural consequence of its exertions. Thus assuming the accomplishment of a rate of motion equal to thirty-five miles an hour to be essential to the successful progress of the balloon, the surfaces by which that result is to be realized must impinge upon the atmosphere with a velocity of not less than fifty miles an hour.

To the sufficiency of this estimate, however, another consideration is necessary; namely, that this impact be maintained at the rate appointed throughout the whole period of the operation : in other words, that no interval or intermission be allowed to take place in the action by which it is generated, whereby the velocity be less at any one period than what is understood to be essential to the accomplishment of the progress required. The necessity for this stipulation will appear the more readily when we consider that the resistance experienced by the balloon is liable to no such periodical suspension ; but, such as it is, may be generally considered as incessant in its operation, at the rate for the time presumed. Whenever, therefore, any cessation or abatement is allowed to take place in the action of the impelling agents, a temporary ascendancy is conferred upon the opposing forces, and a corresponding reduction occasioned in the progress of the machine. To obviate this deficiency and secure a given amount of velocity, in all schemes in which the agents of the propulsion operate by reiterated percussion, (as exemplified in the case of wings or oars) a further accession must be made to the rate at which they are required to act proportioned to the interval allowed to elapse between the periods of absolute impact, and to the difference which that interval is calculated to produce in the momentum of the body, already considered to be fixed at the lowest which the exigencies of the case will permit.

2. When a body is set in motion by the exercise of its agents of propulsion, whatever may be the proportion the resisting surface of the one bears to that of the other, or the difference between the rates at which their impressions are effected, the amount of resistance experienced by each will be invariably the same. Thus, if an individual seated in the car of a balloon ope-

rate upon the adjacent atmosphere simply by means of a lady's fan, a rate of motion, however insensible, will be inevitably induced in the former sufficient to generate an amount of resistance exactly equal to that exerted against the surface of the latter; the only difference being, that in the one case it is concentrated upon a *smaller*, while in the other it is diffused over a *larger* extent of surface, and thus by the infinite participation of its effects escapes detection. From this (which is in fact but a deduction from the well-known maxim of the equality of forces in action and reaction), it follows that to enable the impelling agents to bring up the balloon to a given rate of motion they must be competent to the production of the *same amount* of resistance as the opposing surface or surfaces of the balloon itself, in progress at the rate required.

This result, as we have before had occasion to observe, might (circumstances permitting) be attained equally by a modification of the superfices themselves or of the rate at which their operations are conducted. From what has however been established in the preceding section, I think it will be readily conceded, that admitting even the possibility of the accomplishment of the velocity there assigned to them, we should not be justified in looking to that quarter for any further augmentation by which to enable us to dispense with any portion of the dimensions of the propelling agents which *at that rate* would be necessary to the generation of the required resistance. Now the rate in question being equal, or rather *equivalent* to that of the balloon,* and an equal amount of resistance being required

* It is true that we have assigned a much higher rate of motion to the mechanical agents of the propulsion than that specified as the terminal velocity of the balloon itself; the cause of this distinction, however, being the reduction in the resisting faculty of the medium of the propulsion, occasioned by the natural and necessary condition induced in it by the process itself, in the benefits of which reduction the object to be propelled does not participate, the rates, however different, must be looked upon as the same; being, in fact, only equal to the production of the same amounts of resistance.

as the result of the operation, it follows that the surfaces of the agents by which hat result is to be obtained, must be equal in extent or *equivalent* to those of the balloon itself.

In determining, however, the size of a surface,* by which a given amount of resistance is to be generated at a given rate of motion, regard must be had to the *form* and *structure* of the parts by which the impact in question is to be effected. Upon this head we have already had occasion to comment rather largely in a previous part of this work,† and more slightly in the first section of the present treatise. From what has been there stated, it appears that in creating an impression upon the atmosphere, a plane (and, *a fortiori*, a concave) surface has an advantage over one of a convex or conical construction, varying according to circumstances from one half to one-third of the whole amount. To that extent, therefore, (were there no other obstruction to the progress of the balloon than that arising from the direct impact of the air), might we expect to be able to reduce the proportion between the superficial dimensions of the impelling agents and that of the balloon, in favour of the for-

* The *size* of a surface in relation to its powers of resistance, which is the sense in which it is here used, is a plane equal to the sum of the projections of all the parts by which the progress of the body is impeded, taken at right angles to the line of its propulsion. When the form of the body is simple, this plane resolves itself into a section of the body at its point of greatest lateral extension, or such as its shadow would appear upon a plane surface directly behind it. In bodies of a more complex construction the size of the plane in question is not so easily determined. All parts which project beyond the neighbouring ones, however they may be covered by the intervention of others, receive to a certain extent the impact of the air, and must be considered in the estimate of the surface contemplated in the above definition. The circumstances by which the resistance of the parts so disposed is governed—namely, their relative magnitudes and positions, the degree of shelter they obtain, and the state of the medium when it has reached them—are too vague to permit us to assign any general rule but that of experiment, whereby to determine the exact share they may have in the operation, and how much should be added upon that score to the size of the surfaces by which they are to be matched.

† Letter on the subject of the Parachute. Appendix B. page 230.

mer. From the nature of the proceeding, however,—the complicated construction and extended lateral surfaces of the balloon, modified as it would have to be to suit the purposes of aerial navigation,—a considerable amount of resistance, consisting both of friction and impact, would be developed beyond what a calculation founded upon a consideration of the shape and area of its greatest opposing section would lead us to infer. To meet these accruing demands an augmentation would have to be made in the relative dimensions of the impelling agents, which would remain to be determined by a reference to the particular conditions of the case and the aptitude of the parts to perform the functions allotted to them. Presuming, however, that as far as the *forms* of the parts are concerned, every opportunity would be taken to turn them to the best account —that the surfaces designed to effectuate the resistance, and those whose object it is to evade it would be constructed in the manner most favourable to the interests of each—we may set it down as conclusive that from two-thirds to three-fourths the actual size of the latter would be necessary to enable the former to generate by their impact an equal quantity of a resistance.

In addition, however, to the resistance occasioned by the simple progress of the balloon, there is another obstruction which claims to be provided against by a further extension of the proportions assigned in favour of the surfaces of the impelling agents. This obstruction arises from the necessary opposition experienced by the parts of the latter in the act of recovering their positions, preparatory to the repetition of the stroke by which the propulsion of the balloon is accomplished. Thus, for instance, supposing the machinery employed for the purpose to partake of the nature of the paddles of the steam vessel, rotating upon an axis, while one portion of the apparatus is impinging upon the atmosphere in a direction *favourable* to the end in view, another is inevitably impinging in a direction precisely *opposite*, and with an effective velocity exceeding that of the former by a quantity equal to twice the

the actual rate of the balloon at the time.* This is a conclusion which can never be entirely avoided. No matter how ingeniously devised or how perfectly constructed, a certain amount of surface must ever be presented to the action of the atmosphere in the manner above mentioned, and operate more or less to detract from the value of the forces which it is able to command. How much it would be necessary to extend the dimensions of the impelling agents to counteract the effects arising from this obstruction, experiment alone could accurately enable us to ascertain. As a conclusion, however, which though not expressly deducible from actual calculation is fully warranted by a consideratoin of the case in all its bearings, it may be laid down that, in order to overcome the resistance occasioned by this in conjunction with other obstructions of minor importance, (but which in our general view of the subject it is not necessary at present to particularize), the area of the impelling planes should at least be equal in extent to that of the perpendicular opposing surface of the whole machine itself. Thus, for instance, in the case of the Vauxhall balloon, before quoted, in order that the impelling agents might be able to confer upon it the rate of motion specified as essential to the success of the operation, they must present to the continuous action of the air an extent of plane surface equal, at the least, to sixteen hundred square feet.

In assigning this proportion to the dimensions of the resisting surfaces, it must not be forgotten that much of its sufficiency will depend upon the condition with regard to *continuity* or compactness of the parts by which the impression of resistance is effected. A given extent of surface, distributed into several

* The medium virtually receding as the body advances, the amount of its velocity (= the rate of the balloon) will have to be deducted from that of all bodies proceeding in the *same* direction and superadded to that of those proceeding in a *contrary* one. The sum of these two quantities (= twice the rate of the balloon), constitutes therefore, the difference between the rates of impact of the parts of the machine proceeding in opposite directions.

portions, is by no means productive of the same amount of resistance as if it had been disposed in one uniform plane; neither is one whose contents bear a smaller proportion to the line that bounds them equivalent to one of the same dimensions within the smallest amount of margin by which it is possible to be enclosed. How far again this circumstance would operate to affect the proportion in question, in the absence of actual experiment, can only be conjectured. Regarding, however, the essential conditions of the case, such as we *know* they must be—the uniform bulk of the balloon, and the necessary disjunction of the parts by means of which it is to be impelled—there will be, no doubt, something to add on this score to the allotted dimensions of the latter, which, for the present, we shall only consider as contributing to support the necessity for observing the proportions we have before laid down.

(III.)

To put all this machinery into motion, and confer, as it were, animation upon the lifeless mass, a certain *natural power* is required, the amount and conditions of which it becomes our next duty to investigate. If rapid motion, independent of great force, or great force apart from rapid motion, were all that was sought to be established in the agents of the propulsion, but little difficulty would exist in appointing the means by which it was to be accomplished. By a proper combination of machinery, the smallest conceivable amount of force beyond what is necessary to overcome the inert resistance of the parts themselves, may be so multiplied in its efficiency as to be made to conduce to results in either extremes limited only by the nature of the materials upon which it has to act. A single individual exercising a force equal to one hundred pounds only, by the intervention of a system of six wheels, the circumferences of each bearing to those of their axles the ratio of ten to one, might be made to raise a weight of a hundred million of pounds, or nearly forty-five thousand tons; while, by reversing the action of the apparatus, a rate of motion would be communicated from one extremity of the machinery to the other, a million of times greater than that of the power itself by which it was generated.* It is only where both are cequired to be included in the same operation—where the resistance and the rate, as in the present instance, are both terms of the same proposition—that any limitation exists with regard to the results, or any obligation is imposed upon the means by which they are to be attained.

To those who are acquainted with the principles of the sciences concerned in the case, this conclusion will be at once apparent : for the benefit of others it may be as well to observe that, as it is only by the sacrifice of one or other of the constituent principles of the momentum they are required to communicate (either the velocity or the quantity of matter) that the mechanical powers operate in varying the result of the original impression, whenever a limit is assigned to the extent to which either of these principles may be reduced, a limit is likewise assigned to the advantages the mechanical powers can confer, which draw their influence exclusively from its reduction.

* Supposing the absolute radii of the wheels to have been 10 inches, and those of the axles one inch, then multiplying the former successively into each other, we shall have 10^6=1,000,000 as the value of the leverage in favour of the *power*, and one, (the product of the continued multiplication of the axles), that in favour of the *weight*. Taking then, as above, 100 pounds to be the power of the individual, we have $1 : 1,000,000 : : 100 : 100,000,000$, or somewhat more than 44,642 tons, the weight he would be able to sustain. The velocity, however, being decreased in proportion to the augmentation of the weight, as much as the latter *exceeds* the amount of the original impression, so much will the rate it moves at *fall short* of that of the generating force. In the present case this is a million-fold ; consequently, such will be the difference between the rate of a point in the circumference of the first wheel and that of one in the circumference of the last axle. Supposing, then, the influence of the *power* be suspended or removed, the *weight*, in its preponderance reversing the action of the machinery, would communicate to the *locus* of the former a velocity a million-fold greater than that with which it was, itself, at the time endowed.

To apply these observations to the present question, we have already seen that in order to impel the Vauxhall balloon through the air at a rate of thirty-five miles an hour, a rate of motion in the agents of the propulsion equal to fifty miles an hour is required, generating a resistance equivalent to the weight of 9,528 pounds, or nearly four tons and a quarter. If, instead of this double obligation, it had been simply required to effectuate a resistance equal even to 1000 tons, or a velocity of action amounting to as many miles an hour, the object might easily be accomplished (barring the imperfections of art) by the well-directed efforts of a single individual. As it is, however, no such conclusion is necessary; the mechanical multiplication of the original impression by the sacrifice of the antagonist principles, has already been determined by the appointment of their limits; all further accessions can only be obtained by an *actual* augmentation of its amount. Should the pressure, therefore, which it may be convenient or possible to command, fall short of four tons and a quarter, it must be of such a nature as to develop itself with a rapidity exceeding fifty miles an hour by an amount equivalent to the difference; on the other hand, should the rate of its generation be less than fifty miles an hour, it must exceed four tons and a quarter by a quantity sufficient to compensate the deficiency.

With these facts in view, very little consideration is required to determine the impossibility of effecting the guidance or propulsion of the balloon, to any beneficial extent, by a force originating in the exercise of human strength. This, indeed, is a conclusion which might have been arrived at without any such elaborate computation, by simply reasoning upon grounds deduced from observation and experience; and, indeed, the wonder is, that with so many and such palpable testimonies of the inadequacy of the powers in question, any one should ever have contemplated their employment, or contrived plans, with no more sufficient means to accomplish their execution. Every one who has ever been present at the ascent or descent of a balloon, must have been struck with the display of human force which the occasion is calculated to call forth; the number of men employed in the operation, and the exertions they are compelled to make, at times even when the action of the atmosphere is so slight as otherwise would have escaped their notice. If so many persons, with all the advantages of a solid resting-place, and an unyielding medium for the direct transmission of their strength, can scarcely avail to maintain it in its place, how utterly inefficient must they be when transferred to an unstable fulcrum, and having to apply their force through the intervention of the body itself whose motion it is their object to control?

But the inadequacy of human strength to accomplish the guidance of the balloon is capable of a still more accurate determination. According to the observations of Professor Playfair, Emerson, and others, who have specially investigated the subject, a man of the ordinary powers, working at a wheel, is competent to raise a weight of thirty pounds, through a space of three feet and a half in a second of time, supposing him to continue his exertions for a period of ten hours a day. When the velocity, however, with which he is expected to operate is increased, the amount of resistance against which he can contend must be proportionately diminished; and, at the rate ascribed to the agents of aerial propulsion, (namely, fifty miles an hour, or seventy-three feet in a second,) could only be estimated at about one pound and a half;* that is, presuming him capable, at the rate in question, of overcoming the *vis inertiæ* and friction of the machinery with which he would have to contend.

---

* Seventy-three feet in a second (the rate required) being *twenty-one* times greater than that contained in the proposition upon which our estimate is founded, the weight which could be raised will be but a *one-and-twentieth* part of that referred to in the same proposition. Thirty, divided by twenty-one, gives very nearly the quotient we have above deduced.

By the substitution, however, of his legs instead of his arms, a higher degree of power might undoubtedly be obtained, and which might be still further increased were he accommodated with such an apparatus above his shoulders as would enable him to add some amount of muscular reaction downwards to that accruing from the sole exercise of his bodily weight. By this means, at the ordinary rate of walking, (which may be roughly assumed at three miles an hour,* or somewhat more than four feet in a second,) a man might, for a considerable length of time, exert a force equivalent to his whole weight, or about one hundred and fifty pounds; which reduced in proportion to the increase in the rate (namely from four to seventy-three feet in a second, or about eighteen-fold) would give a result of eight pounds nearly as the available extent of each individual's exertions.

The exercise of muscular strength, however, no matter how lightly it may be taxed, being limited in its duration, while the estimate upon which its amount has been determined is founded upon the supposition of its uninterrupted continuance, it would be necessary to be provided with such an amount in reserve as would suffice to maintain the same quantity of power in constant operation. Admitting, therefore, that a man could continue to work at the rate ascribed to him for one-half of his time, a double supply of men, at the least, would be absolutely requisite; whereby the amount assignable to each individual would in effect be reduced to only four pounds; a quantity bearing so small a proportion to the weight as to hold out no prospect of its ever being available in the practice of an art, the main condition of which is the attainment of extreme specific lightness. To illustrate this conclusion by reference to a particular case, we have already seen that the resistance experienced

* A man may walk at the rate of four miles an hour, but I doubt if he could exercise his legs in the mode which would be required in turning a wheel, with the same freedom and at the same rate as if he had merely a progressive motion to accomplish.

by the Vauxhall balloon in passing through the air at the rate of thirty-five miles an hour would be equal to 9,528 pounds, or about 2,400 times the amount of that ascribed to each individual; consequently to effect its propulsion consistent with the obligations we have already considered to be essential to the accomplishment of any beneficial result, would require a force of 2,400 men, or about two hundred times as many as her whole ascensive power would be competent to support; and *that*, too, making no allowance whatever for the weight of the machinery by which they would have to operate.

It is true, by the adoption of another form, a balloon requiring no more propulsive power than that we have made the subject of the preceding calculation, might be constructed capable of supporting four times the weight: even here, however, all that would be effected would be an increase to that extent in the efficiency of the cargo, which would still remain about fifty times as great as she would be able to support.

Nor is this a conclusion which could be avoided by *reducing* the size of the balloon, in the hopes of attaining a point in which the forces opposing and those opposed would be more on a par. On the contrary, the resistance varying as the *squares* while the buoyant power follows the ratio of the *cubes* of the diameter, any attempt to diminish the scale of the experiment but tends to magnify the disproportion between the difficulties and the means whereby they are to be encountered; an elliptical balloon of nine feet radius, equivalent only to a charge of two men, (the smallest number consistent with what we have before stated to be necessary for the due continuance of the impression), developing at the rate in question a resistance of 1,024 pounds, and consequently requiring an amount of human power at the value we have assigned to it, one hundred and twenty-five times as great as it is capable of raising. For the satisfaction of those who might expect a more favorable result, by *enlarging* the dimensions of the balloon, we have subjoined a calculation from which they will perceive that, in accordance with the obligations before laid down, the smallest number of men that could propel a balloon sufficient to support them would be about three millions three hundred and thirty-five thousand, and the smallest balloon that could carry men sufficient to propel her at the rate in question would be equivalent in its contents to a sphere of about three thousand two hundred and sixty-three feet in diameter.[*]

[*] The following is a general formula for calculating the direct resistance upon all balloons, partaking of the nature of a sphere, cone, cylinder, or ellipsis:—Square the radius of the largest section perpendicular to the horizontal axis of the machine, and multiply by the circumference expressed in terms of the diameter;[*] this gives the number of square feet in a circular plane equivalent to the said section. Of this, two-thirds only are to be considered as forming the real amount of the resisting plane, (the actual resistance being upon an average diminished one-third, on account of the particular form of the opposing surface); which multiply by the sum answering to the rate of the wind in the table of atmospheric resistance, and the product will be the amount of direct resistance in pounds avoirdupois. Divide this sum by the number of pounds which, at the rate assigned to the agents of the propulsion, shall be found equivalent to each man's muscular strength, and double the quotient will represent the number of men required to effectuate the same amount of resistance at the same rate, supposing one change of men sufficient to perpetuate the operation. This formula may be algebraically expressed as follows: $2\frac{\pi r^2}{3} \times w = W$, where w is the resistance in pounds upon each square foot, and W the effective resistance upon the whole surface: $2\frac{W}{P}$ gives the number of men required, P representing the amount of force which each individual can bring to bear at the rate assigned to the agents of the propulsion.

By this mode of computation may be tested the conclusion we have arrived at in the text. As the buoyant power of the balloon follows the ratio of the cubes, while the superficies, and consequently the

[*] This mode of expression is perhaps too elaborate for the general reader, and may therefore require explanation. The "circumference, in terms of the diameter," implies the *proportion* between these two parts of a circle, or the quotient obtained by the division of the latter into the former. This proportion, not being exactly determinable in numbers, can only be represented by an approximation, of which the sum 3·1415, in decimals, is sufficient for most practical purposes. In Algebra, this proportion is expressed by the Greek letter Π: as in the formula Πr^2, signifying the superficial contents of a circle, obtained by multiplying the radius squared (r^2) into the sum expressing the proportion in question.

In default of human strength, the mind naturally reverts to the great agent of modern invention, the wonder-working power of steam. Independently, however, of the inconvenience and danger necessarily attendant upon the employment of a power requiring the aid of fire, there is one essential objection to steam which must for ever preclude the possibility of its adoption as an agent in the propulsion of the balloon; I mean the continual *loss of weight* from the consumption of fuel and the conversion of water into vapour, which more or less must ever attend its employment. The force of this objection will at once

resistance, varies as the squares of the diameters, it follows that any alteration in the size of the balloon must affect the former more than the latter; if a balloon, therefore, is capable of carrying *exactly* the quantity of human power equivalent to the resistance she develops, she must be the *smallest* that can be constructed with such a result; inasmuch as any further reduction in her size would diminish her buoyancy more than her resistance, and she would then require more force to her propulsion than she would be able to carry. Now, considering a balloon of fifty feet in diameter, when properly inflated, to be sufficient to raise a weight equivalent to twelve men, by referring to the proportion between the cubes of their diameters, we shall find that one of three thousand two hundred and sixty-three feet, quoted in the text, would be barely competent to a charge of 3,335,204 men. By throwing the gaseous contents, however, into a more elongated form, it would be possible, as as we before observed, to reduce the resistance without affecting the buoyancy. Such a vessel would be a cylinder, capped with cones, or an ellipsoid, whose tranverse axis was two thousand and fifty feet, and length equal to four times its diameter. The resistance occasioned by the direct *impact* of such a body in progress through the atmosphere at the rate of thirty-five miles an hour would, accordingly, (as will be seen by reference to the preceding formula) be equal to 13,337,160,354 pounds, and 3,334,290, the number of men by which an equal amount of force could be generated; each man's quotum being eight pounds, as above assigned, and a double allowance of men being required to aumit of the operation being carried on without interruption. The difference (amounting to nine hundred and fourteen) between the number of men equivalent to her resistance and that equivalent to her buoyancy, as here displayed in favour of the latter, however *less*, would certainly not be *more* than enough to compensate for the weight and resistance of the machinery, the friction of the atmosphere, and other circumstances, more or less influential, which have not been included in the above calculation.

appear, when we consider that it is by the preservation of the equilibrium between her contents of gas and ballast she maintains her position in the air. Whenever that equilibrium is disturbed by the abstraction of a part of either of these resources a sacrifice of a proportionate amount of the other becomes absolutely necessary in order to restore it; a proceeding, it is scarcely necessary to remark, by which her whole efficiency must sooner or later become destroyed. This objection equally applies to all those powers which are obtained by means of chemical decomposition, the rapid generation of gases by explosion, combustion, or otherwise; the very efficiency of which is, in fact, only commensurate with the loss of weight by which they are accompanied; nor, indeed, am I aware of any whatever, applicable to the purposes in question, unless indeed it may be that of electro-magnetism, concerning which, however, our information is yet too limited to allow us to speak more decidedly.

(IV.)

Possessed of these, the mechanical agents of the propulsion, together with a power sufficient to invest them with motion at the rate and under the development of pressure before calculated, the aerial engineer must not conclude that the question of the guidance of the balloon has been completely solved, and that nothing remains to interfere with its immediate adoption as a mode of transport applicable to the ordinary purposes of life.

Independent of the difficulty that must ever attend the reduction to practice of rules involving the nicest points in rational and practical mechanics, the most rigorous economy of power, and an intimate knowledge of the strength of materials, with the best method of employing them, there are certain restrictions regarding their application, failing compliance with which the best-devised schemes for the propulsion of the balloon must prove utterly inefficacious, or at least successful to so small an extent as to remain still as inapplicable as ever to the purposes for which they are required.

The first of these regards the form of the aerial vessel. It is scarcely necessary to observe, that before any scheme for its guidance be attempted, the balloon itself must be of such a form as will admit of its being guided. It must have a line of least resistance, and this line must be that, in the direction of which it advances. This involves, likewise, the consideration of a rudder, or some other corresponding apparatus, by means of which its propulsive energies may be directed into a determined channel. In short, it must have a head and a tail, as well as a body, and be capable of assuming and maintaining a fixed position during its forced progress through the air. Such a form, for instance, would be that of an ellipsoid, as before observed, or a cylinder terminated by cones, like that recently exhibited to the public by Count Lennox, under the name of the aerial ship, and of which representations are to be found in old prints of aerostation, illustrative of previous projects for the guidance of the balloon.

In the second place, it must be so contrived that when subjected to the action of a strong current of air, the balloon shall not, in the change of position it will be inevitably forced to adopt, interfere with the action of the machinery by which it is impelled. In regard of this, as indeed of all the other rules, consideration must be had, not to the *actual* shape and position of the balloon, but to that which it will have assumed when acting under the influence of the opposing forces.

Thirdly, it follows from this, as a matter of course, that the same strength of materials which is found sufficient for an ordinary balloon, would by no means suffice for one, the nature of whose employment infers the exposure to excessive and unwonted opposition.*

* Of the necessity for this provision the French projectors seem fully sensible, when they advert to the possibility of forming the balloon itself of solid materials, and gravely look forward to the time when wood, copper, iron, and the other ingredients of terrestrial and marine architecture, shall be put in requisition to supply a more substantial vehicle for the occupation of the empty regions of the sky. Upon the practica-

Fourthly, the whole must be so constructed as not to suffer from the shocks to which it will be unavoidably subjected whenever it comes into contact with the ground, owing to the impossibility of making the attachment to the earth with that degree of firmness and certainty which is necessary to ensure the safety of the balloon and place it under the immediate control of the aeronaut. And this, it strikes me, is one of (if not actually) the most important of the practical restrictions in question, and, at the same time, the most difficult to be complied with, consistently with the other essential features of the case. For what, after all, can be the merit of any machinery that is liable, nay, almost certain, to be rendered valueless whenever it may happen to be employed, except under such a favourable juncture of circumstances as is not to be counted upon in the practice of an art, carried on under the auspices of proverbially the most fickle power in nature? And, yet I must confess, I do not see any means of avoiding this conclusion by any structure of machinery that shall be in accordance with the rules we have before laid down for its regulation. The great extent of surface, and the lightness which ought to be its primary characteristic, are qualities equally calculated to aggravate the effects of the opposing forces, as incompatible with the requisitions of strength by which alone they could be successfully resisted; and, indeed, it is difficult to conceive any structure or arrangement of machinery suitable to the purpose, that shall either be beyond the reach of the violence to be apprehended, or sufficiently strong to avoid suffering essential detriment from it, whenever it occurs.*

bility of such schemes, it would be useless to waste words; I should only like to know, when formed, how it is to be inflated, and when inflated how it is to be emptied; for it is not to be forgotten that before it can be inflated it must first be emptied, while, at the same time, once it is filled, nothing can be abstracted from it without the introduction of an equivalent. This latter consideration would, I rather suspect, leave the office of the valve somewhat in the nature of a sinecure.

* The disregard of this particular constitutes one of the most remark-

Fifthly, the agents of the propulsion must be made to operate directly upon the body of the balloon itself, and not, as in every scheme heretofore projected, upon the car which is attached to it. In the fulfilment of this condition a great difficulty presents itself in the different nature of the materials which will have to be employed in the construction of the balloon and of its machinery; the flexible quality of the one, the solid unyielding nature of the other, and the certain danger to the former when united firmly to the latter under

able characteristics of all the aerial projectors with whom I have ever communicated. Treating the balloon merely as a manequin, to *try on* schemes of propulsion, they entirely neglect to consider the condition it will be placed in when it comes to be exposed to the influence of the forces it will have developed in its career. Hence the inefficacy and absurdity of most of their contrivances whenever any attempt has been made to reduce them to practice. One of the adjuncts to the original plan of Count Lennox's air-ship was, I remember, a set of small wheels fastened beneath the car, (or rather the canoe) to the frame of which the motive agents were to be appended, in order to enable it to glide on the earth after the descent, and avoid the consequences of a too sudden interruption to its flight! Imagine a piece of machinery sixty feet broad and one hundred and eighty long, bearing a charge of more than ten tons, and furnished with wings projecting some forty feet or more on either side, gliding over the country upon castors, under the influence of a wind moving at the rate of thirty or forty miles an hour, attached, *for steadiness*, to a vessel of still more preposterous dimensions, floating over head and exposing to the action of the wind an extent of surface equivalent to upwards of twenty thousand square feet! Indeed the speculative Frenchman seems to have entertained a strange notion of the nature of the element with which he was about to contend, when, in reply to the suggestion of a gentleman concerning the security of his machinery in the descent, he observed that it would be easy to obviate all danger upon that score by coming down under the lee of some building or high wall, by which he would at all times be sure of being properly sheltered from the wind! an ingenious expedient, as Mr. Green slyly observed, which might be considerably improved upon by the addition to his cargo of a *ready-made north-wall*, suited to all cases of emergency; upon the principle, no doubt, of the *universal finger-post* which the Irishman sagaciously proposed to the celebrated African traveller, Captain Clapperton, as a ready means of solving his doubts whenever he should happen to have the misfortune of losing his way in the deserts!

exposure to forces such as may be expected to accompany the operation of aerial propulsion.

Sixthly, the construction of the machinery must be such that an injury to one part shall not necessarily impede or prevent the action of the rest, or be attended with consequences involving the *security* of the balloon.

And, lastly, though not least, the whole must be so contrived as to maintain its equilibrium under all the variations of force to which it will be inevitably subjected in its progress.

These then constitute the principal obligations which the nature of the proceeding has imposed upon the guidance of the balloon. From a consideration of what has been discussed in the preceding sections, the ingenious reader will, no doubt, have observed that the main obstacles to the accomplishment of the object in view are, first, the construction of surfaces of the proper degree of lightness, and of sufficient size and strength united, to enable them at once to *effectuate* and *withstand* the pressure they are required to afford ; secondly, the adaptation of a power competent to invest them with the proper motion ; and, thirdly, the arrangement of the whole machine in accordance with the principles laid down in the latter section.

A fourth obligation, however, of equal, if not superior importance to any, yet remains to be commented upon ; namely, the regulation of the motive agents in such a manner as to ensure by their impact the resistance which has been assigned to their operation. The difficulty of complying with this requisition is one proceeding from the elastic nature of the medium, whereby its equilibrium of density becomes more easily disturbed, and a state of rarefaction induced in the portions contiguous to the surfaces in question, to the manifest deterioration of the resistance they are expected to create. This will be better understood when we consider, that upon the rapid passage of the surfaces in question a large portion of the adjacent atmosphere is swept away in the direction of their impact, leaving throughout their whole course a medium more or less rarefied in proportion to the rapidity with which they operate. To this result both the rate and size of the moving planes essentially contribute ; and there is no doubt that long before either of these conditions were fulfilled to the extent assigned in the estimate of their respective quantities, a considerable approximation to a vacuum would have been formed in the locus of their operations, requiring more or less time to fill up, in proportion to the extent of space it had affected. Now should it happen that the planes in question be compelled to reiterate their percussion within the sphere of this disturbance ere the atmosphere has had time to recover from its effects, a drawback to their efficiency will be occasioned which no increase of rate or dimension will enable them entirely to overcome.

Upon the whole review of the case then it must be avowed that the propulsion of the balloon to the extent we have imposed upon it, is beset with difficulties of no ordinary description. It is true that these difficulties consist not so much in the *quality* as in the *quantity* of what is sought to be done—in the *nature* of the operation, as in the *extent* to which it is requisite that it should be accomplished. Hence the possibility of effecting in a minor degree that, to which considerations of paramount importance have induced us to assign a more extended limit. Apart from other considerations, the question of the guidance of the balloon is a mere expression, conveying no definite idea and affording no certain grounds for investigation. As a mere abstract fact, there is no doubt the balloon can be guided; it is only in reference to the particulars of the case, that any question can arise upon the matter. When, therefore, any person says that he has discovered the means of guiding the balloon, his assertion literally amounts to nothing, unless, at the same time, it be coupled with a specification of the rate and conditions under which he is able to effect it. Should these be found to correspond with what has been stated in the preceding sections,

then, and not otherwise, will the question of an aerial navigation, applicable to useful purposes, have been duly and satisfactorily determined. This, however, is a consummation which
I fear there is but little prospect of our ever being able to
attain. The deficiency of power and the limitation assigned
by nature to the strength of materials, form a barrier which all
our efforts seem incapable of enabling us to surmount; and,
indeed, when we consider the nature and amount of the forces
required to the propulsion of the balloon, it becomes a matter
of question whether the same exertions would not be sufficient
to enable us to dispense with its services altogether, and transport ourselves through the air by the simple exercise of wings
alone.*

The reader must not be misled by those insidious analogies
by which unreflecting persons are wont to be guided in their
sentiments upon matters of this description, nor conclude that,
because a ship sails, a fish swims, or a bird flies, it is equally
consistent with the laws of nature that a man should be able to
direct his course through the atmosphere by the aid of a balloon. Ample reasons will be found in the circumstances of each

* The reader may not perhaps be aware that the bold idea of human
flight has once to a certain extent been actually realized, and that one
individual, almost within the memory of man, has been known to raise
and conduct himself through the air by the agency of wings alone.
The instance alluded to is that of the Marquis de Bacqueville, who in
the year 1742, according to a notification which he had made to that
effect, rose in the sight of the assembled multitudes of Paris, from
his own residence on the *Quai des Theatins*, and directed his course
across the Seine towards the gardens of the Tuilleries, whither he had
signified his intention of proceeding. At first he appeared to advance
with tolerable steadiness and facility; when about half-way over,
however, something occurred which has never been thoroughly comprehended, by which he seems to have been deprived of the power of
continuing his exertions; when his wings ceasing to act in the manner
necessary for his support, he sank to the ground and was precipitated
against one of the floating machines belonging to the Parisian laun-
dresses, which line the arches of the *Pont Royale* on the side of the
river opposite to that from which he had taken his departure, whereby
his leg was broke and other serious injuries inflicted upon his person.

to invalidate these analogies and disprove any dependance which might be conceived to exist between them. The ship commands her course over the bosom of the ocean, not from the simple fact alone of her possessing two elements endowed with different rates and inclinations of motion, (for such a reason would exclude the steam-vessel from our argument, which secures her progress by the instrumentality of one alone), but also from the striking superiority in the density of that (the water) to which she resorts for her propulsion over that (the air) in which so large a proportion of her mass is destined to move;* while, at the same time, the general condition of the former, as far as its progressive motion is concerned, is such as to require but a comparatively moderate share of power to enable her to contend with it. Of these the latter is an advantage equally enjoyed by the finny inhabitants of the deep; and though it is true the former (namely, a difference of density in favour of the medium of propulsion) does not characterize their condition any more than it does that of the balloon, yet the want of it is more than compensated by the possession of a specific gravity, so nearly on a par with that of the element in which they move, that little or no accession of bulk is required to enable them to support themselves that does not likewise contribute to the enhancement of the strength, by which they direct their motions. The example of the bird, it is true, appears at first sight to be more to the point: possessed as it is of a specific gravity scarcely more favourable to its support than our own; while, at the same time, the medium of its evolutions being the same as that of the balloon, the same impediments remain to be encountered by them both. The analogy, however, although certainly more specious than the proceeding, is by no means more conclusive. For both these emergencies nature has supplied a remedy; for the former, in the endowment of immense

---

* A reference to the operation of the motive agents (Page 328.) will show that the very reverse of this is the relative condition of the medium of propulsion and that of opposition, in the case of the balloon.

muscular strength; for the latter, in the *actual* smallness of their dimensions. Possessed of a power sufficient of itself to overcome the attraction of gravitation, the efficiency of the animal is ever dependent upon its bulk, and consequently at all times proportioned to the resistance it has to contend with; while from the *positive* smallness of its size, not only does the structure of its organs easily fall within the limits assigned by nature to the strength of the appropriate materials,* (in consequence of which it is enabled to surmount a great portion of the forces arrayed against it, and at all events avoid incurring damage from the remainder), but likewise through the facility with which it can secure a retreat, it is enabled without prejudice to dispense with the possession of powers superior to what at times it may have occasion to encounter : confined to minute dimensions, the bird that is unable to match with the wind can at every turn find a refuge from its influence, and is consequently perfect with half the comparative amount of

* Even if a man were endowed with the same proportion of muscular strength as a bird, with the same natural organization to enable him to apply it, he could not, for the reason here mentioned, ever turn it to the same account; inasmuch as, with all his powers to fly, he could never procure material that would admit of sufficient extension to construct the organs of his flight. Hence, man, though he may succeed to a certain extent, as in the case of the Marquis of Bacqueville, will never be able to dispute with the feathered tribes the empire of the sky; not because he could not *exercise* his wings, but because he could not *make* them. Of the limitation thus imposed by nature, the strongest and most striking examples are afforded in the works of nature herself: when the birds of *her* creation exceed a certain size they do not fly. It is not because they are heavier in proportion to the density of the medium, and therefore want the requisite degree of muscular strength; for that *is* in many cases, and *could* in all, be supplied by nature without any infraction of her existing laws; it is because Providence has not thought proper to create a material adequate to the construction of their organs. The emu, the cassowary, the dodo, the ostrich, are birds in all but the possession of wings; may we not conclude that, if the materials for their construction had already existed, nature would not have left the noblest specimens of her work imperfect? And, can man hope to succeed, where nature has declared her inability to prevail?

force which would necessary to the success and security of the balloon. Thus to sum up,—a density in the opposing medium inferior to that of the medium of propulsion—a specific gravity but slightly removed from that of the element in which they move, together with comparatively trifling forces to contend with—and, lastly, a size that arms them against injury and puts security at all times within their reach. These are advantages more or less enjoyed by all objects affecting fluid media which are denied by nature to man in his endeavours to navigate the atmosphere, and completely destroy whatever analogy might be thought to exist between them.

"But," it may be asked, "supposing us unable to accomplish all that has been stated to be necessary to the perfect government of the sky, why may not an aerial navigation be made applicable to useful purposes in a *less* degree? And why must we abandon *all* hopes of advantage from the practise of an art because we are unable to bring it to a higher degree of perfection?" Simply because, in reducing the rate, (upon which hinge all the essential difficulties of the case), we sacrifice altogether that condition by which the character of an art, *applicable to useful purposes*, is essentially distinguished; namely, the certain prospect of success. It is not that the object would be accomplished with less speed, less safety, or to a less extent; but that in adapting our resources to a scale of opposition inferior to what we may have to encounter, we forego the *certainty* of ever accomplishing it at all. For all purposes, where this condition is a matter of indifference, an aerial navigation might no doubt be established; but as it is this "regard to the result," that, as I take it, constitutes the main difference between affairs of business and affairs of pleasure, still would the latter alone have all the benefit of our exertions.

APPENDIX E.

ON THE MEANS OF MAINTAINING THE EQUILIBRIUM OF THE BALLOON.

THE power of altering the elevation of the balloon without the expenditure of gas or ballast, and consequently enabling her to continue for a longer period in the air, is an important adjunct to the practice of aerostation, which from the first period of its discovery it has constantly exercised the ingenuity of its votaries to supply. And yet, simple as the circumstance may appear, none of the various schemes which have hitherto been suggested for the purpose, (with the exception of Mr. Green's, just described), has ever been found to conduce in the slightest degree towards the end for which it was designed. Of these, indeed, many are so extravagant as almost to induce a doubt whether they were ever seriously intended to be applied; such, for instance, is that which proposed to operate by condensing the aqueous vapour in the sky to procure an accession of ballast; and still more, (if, in truth, there can be *degrees* in utter absurdity) the project of maintaining the supply of gas by the chemical decomposition of rain or other watery deposit, which they expected to be able to obtain during the continuance of the voyage; and, indeed, no better illustration can be afforded of the extravagance of most of these devices than the fact, that it has baffled the ingenuity even of their own projectors to submit them to the test of experiment.

The first plan for the purpose that was ever actually tried was that of M. Pilâtre de Rosier, subsequently adopted by Zambeccari and others, and consisted of a combination of the ordinary hydrogen-gas balloon with one on the principle of M. Montgolfier, by acting upon which latter, either increasing or diminishing the density of its contents, they conceived the hope of affecting the equilibrium of the whole. The fatal results of their experiments to the former aeronauts, with the occurrence of accidents more or less serious to the rest of those who adopted it, soon however put a stop to the further employment of a scheme which, however ingenious in theory, was too dangerous in practice to be made available to an art, already, it was thought, sufficiently so without it.

The next plan deserving of notice is that of M. Charles, which he attempted to put in execution in the excursion from St. Cloud, in company with the Duke de Chartres, M. Robert, and another gentleman. The mode by which it was intended to operate in this case was by the condensation of atmospheric air in a vessel contained within the body of the balloon; owing, however, to an accident which occurred during the preparations for the ascent, the vessel became dislodged from its bearings, and, falling over the neck of the balloon, frustrated all their endeavours to reinstate it in its proper position, or to make use of it in the manner they had intended. No absolute result, therefore, was ever obtained from this experiment, nor has it since been repeated; no doubt, from a conviction of its insufficiency in the minds of all practical aeronauts. But, indeed, this conclusion might have been arrived at, without having recourse to any such experiment at all. The weight, *in vacuo*, of one cubic foot of atmospheric air, taken at mean pressure (thirty inches) and temperature 55°, is pretty accurately one ounce and one-fifth; consequently, the quantity of the element that would be equivalent to a pound in weight, under the same conditions would occupy a space of thirteen cubic feet and one-third. The weight of all bodies suspended in fluid media being, however, diminished by an amount equal to that of a like bulk of the medium itself, the actual gain by the process of condensation is, really, only equal to the difference between the intrinsic weight of the whole quantity condensed, and that of the original contents of the recipient vessel, at the pressure of the atmosphere of the place, and for the time being. That a bulk, therefore, of thirteen cubic feet and one-third should produce the effect of one pound weight, its density must be made double of that of the external air, or as it is technically phrased, it must be submitted to a pressure of two atmospheres; if a weight of two pounds be required of the same bulk, it must be compressed with a force of three atmospheres; if three pounds, four atmospheres, and so on; the weight increasing directly with the pressure, and the value of one measure of the vessel being lost in effect, from the influence of the medium in which it is suspended. To this process of condensation, however. there is a limit in practice which is more or less speedily attained, according to the facilities which are afforded for the conduct of the operation. Considering how much these are restricted by the nature and circumstances of aerostation, I certainly think we should not be justified in calculating upon a further extension of the process than is equivalent to a pressure of *five* atmospheres, so that, in fact, the utmost extent to which a bulk of thirteen cubic feet and one-third, could by these means be rendered available for the purposes of ballast would be equal only to about four pounds; consequently, to accomplish an alteration of four hundred pounds in the equilibrium of the balloon, (and, from what has been observed regarding the influence of the nocturnal moisture,* less would not be sufficient for all emergencies) it would be necessary to be provided with a vessel capable of containing at least one hundred times the quantity alluded to, or one thousand three hundred and thirty-three cubic feet, which is equivalent to a sphere of about fourteen feet in diameter. If to the preposterous bulk and consequent weight of the apparatus, we add the complicated machinery by which it would have to be worked and the length of time it would require to accomplish the operation, it will appear very questionable whether any advantage could ever be made to accrue from the proceeding in question, sufficient to warrant its adoption.

* See page 14.

The same arguments which we have here adduced apply with still greater force in regard of any attempt to operate by condensation upon the gaseous contents of the balloon itself. This will be best illustrated by observing the further augmentation in the amount which would be required to produce the same effect according to this latter mode of proceeding. The gas which is generally procured from the manufactories for the purpose of inflation, has about one-sixth of the specific gravity of atmospheric air; consequently, to destroy altogether the buoyant property of any portion of it, and simply neutralize its influence upon the balloon, it would be necessary to increase its density six-fold, by forcing it to occupy one-sixth of its natural volume. Of this gas, about fifteen cubic feet may be taken to be adequate to the support of one pound weight: hence, to effectuate a difference of four hundred pounds in the power of the balloon, it would be necessary to subject to a compression of six atmospheres, a volume of gas equal to six thousand cubic feet; about six hundred and sixty-seven cubic feet more than would be required to be operated upon by the preceding plan; notwithstanding we have considered it right to limit the condensation in the former instance to five atmospheres, or one less than what we have assumed in the latter.

A more simple method of preventing the variations in the equilibrium of the balloon, might perhaps be founded upon the exercise of large resisting planes, or wings, capable of operating in a vertical direction, and so constructed as, at all times when in motion, to present one pair, or set of surfaces, to the action of the air, while the others are recovering their pristine positions, and preparing for a renewal of their impact. The effect of this operation would, of course, depend upon the size of the surfaces, and the power by which they were called into action; it is, however, to be observed that the main obstacles to the successful guidance of the balloon by means of wings, upon which we have commented in the preceding Appendix, do not operate to the same extent in the present instance; both the forces themselves being incomparably less, and no restriction of *rate* existing to determine the amount of power by which they are to be generated. With all these advantages, however, it is much to be doubted whether any operation, requiring a complicated machinery and the constant exercise of manual force, will ever be made available to any beneficial extent; nor, indeed, does there appear, out of the many schemes which have been suggested for the purpose, any that have not some inherent practical deficiency that, even were they otherwise unobjectionable, would not render them inapplicable to the purpose for which they are required. To supply, at an instant's notice, a compensation for any accession of weight which it is possible to expect, without attention or effort on the part of the aeronaut, inferring no risk, and procurable at a trifling expense, are qualities confined entirely, as yet, to the simple plan of Mr. Charles Green.

Having already entered at large into the merits of this discovery, in the introduction to the previous narrative, it only remains to advert to one circumstance, of which, as it was not necessary to the illustration of the subject in connexion with the expedition, we thought proper to remit the consideration to the present opportunity.

From the circumstance of the inapplicability of either of the devices there detailed to the element for which it was not particularly intended, (namely the incapability of the simple guide-rope, from its sinking, to afford the necessary relief when over water, and the impracticability of exercising the water-ballast over the surface of the earth), it will, no doubt, have struck the reader that, in cases requiring the adoption of either arrangement in succession, (as where the course of the balloon happens to embrace a varied tract of land and water), the instrument, as it stands at present, must fail in producing the desired effect; inasmuch as, when the first tract of sea had been passed and the balloon was about to enter upon the land, it would be necessary to discharge ballast in order to carry her and her floating appendages clear of the coast, preparatory to having recourse to the simple guide-rope; and the quantity of ballast required for this purpose being precisely the same that it would have taken, in the first instance to have prevented the depression, all the advantages that appeared to have been secured in the interim would be forfeited at once, and the balloon reduced to exactly the same condition, as regards her resources, as if the expedient in question had never been resorted to at all.

This defect, arising from the impossibility of suspending the action of the instrument for a time, in order to make the required alteration in its condition suitable to its new situation, is evidently to be obviated only by the substitution, for the *two* systems, of *one* that shall have the power of accommodating itself to both contingencies. Were it possible to construct the guide-rope of materials specifically lighter than water this object to a certain extent would be accomplished. As, however, this does not appear at present to be possible, and would besides, if adopted, essentially detract from its efficiency, when exercised over its own appropriate element, (the land), the method by which I propose to remedy the defect is to attach to the guide-rope, immediately upon reaching the sea, a certain number of pieces of cork, such as are used for fishing-nets, (or, otherwise, simple bladders of air,) sufficient to prevent its submersion when over the water, but calculated to give way whenever they begin to encounter the more solid opposition of the earth. Should the depression in the elevation of the balloon, which calls for the intervention of the guide-rope, have occurred when in the act of traversing the sea, the arrangement in question can be easily accomplished in the first instance, while delivering the guide-rope from the windlass. Should, however, the case be otherwise, and the progress of the balloon, under the influence of the disturbing causes, have called for the employment of the guide-rope before the land be passed, it will be necessary upon reaching the sea to draw in the rope in order to make the proposed attachments; during which operation a sufficient quantity of the usual water-ballast is to be suspended from the car or hoop at a distance of about twenty or thirty yards below; by which means the depression of the balloon, (consequent upon the retraction of the rope in the one case or its sinking in the other), will be arrested and an opportunity afforded to complete the necessary arrangements; upon the conclusion of which, the guide-rope is again uncoiled and, lightening the balloon as it runs off, allows her to resume her original elevation, equally adapted for either element, and capable of repeating the operation as often as occasion may require.

APPENDIX F.

As an adjunct to a work, professedly designed to illustrate
the navigation of the air, I have thought that a few observations
upon the proceedings of an animal frequenting the same element might not prove uninteresting to the reader addicted to
such inquiries. The subject, indeed, as far as the simple operation of flying is concerned, has already been pretty largely
discussed by most ornithologists of distinction, and more especially by the renowned Alfonso Borelli in his investigations
" De Motu Animalium," published in 1680 and 1681.

Notwithstanding, however, the frequency with which it has
been treated, and its own apparent simplicity, still it does not
appear to me that the principles upon which it is conducted
have ever been so fully explained as to convey a clear and definite idea of the *modus operandi*—the precise nature of the efforts
by which the feat is accomplished. That a bird elevates itself
into the bosom of the atmosphere by the reaction of its wings
against the subjacent air, is a fact we all know and can comprehend without other reference than to the simplest principles of
mechanics. This, however, is but a small part of the operation
of flying, and one which, with a little contrivance, we might be
enabled to execute with even still greater facility ourselves.*

* If the mere circumstance of elevating himself into the air, by the
mechanical exercise of wings were all that was necessary to enable a
man to fly, the only obstacle to this conclusion, (being his own weight,
and that of his machinery), would be easily overcome by the addition of
a balloon of the requisite dimensions, so as to reduce the weight of the
whole to such an amount as his existing resources could easily enable
him to compete with. As there is no limitation to the extent to which

It is in the direction of this flight, the multifarious and complicated evolutions by which the course of the bird is distinguished, that nature evinces the superiority of her work, and distances our comprehension almost as completely as it defies our art. What, for instance, can be more wonderful, as a specimen of locomotion, than the rapid and continuous passage of a bird through the branches of a tree, without the slightest apparent confusion, derangement, or interruption to its course, at a time when, perhaps, the very medium of its support is itself moving with equal rapidity in another direction ; and how inferior in his prowess is man himself with all his boasted attributes, that can scarcely walk, much less run, through a scattered grove of trees, or across a street encumbered with carriages, without either actually falling foul of them in his career, or at least evincing by his hesitation and the awkwardness of his gait, the fear he entertains of such a conclusion, and the difficulties he experiences in his efforts to avoid it. If such be the difference between them when each enjoys the benefit of his own appropriate element, how much greater must it appear in favour of the former, when the latter consents to relinquish his native sphere and transfers his endeavours to the arena of his feathered rival?

The variety and delicacy of the physical exertions by which the bird is enabled to execute the evolutions ascribed to it are, indeed, truly astonishing ; and, when we compare its condition in that respect with the clumsy efforts of man's most successful imitations, we cease at once to marvel at his failure or at their success. In the one case, we have a body composed of a variety of parts, every one of which is endowed with vitality,

he could reduce his weight, so there is no question that his efforts to raise himself must be ultimately successful. The size of the smallest balloon, however, which could be made available to the purpose in the slightest degree, would still be so great as to overpower all his efforts for its guidance, to nearly the same extent as if he had made it large enough to support the whole.

susceptible of distinct and separate motions, obedient to the slightest impulse of the will, constructed of materials made expressly for the purpose, and from constant habit and natural aptitude, capable of executing the most complicated movements with a security, efficacy, and despatch, that seems almost to anticipate the inclination by which they are ordained : in the other, could we even accomplish his equipment and give him power to avail himself of his means, we should still have an object composed of *substitutes* instead of *attributes*—a heterogeneous mixture of remedies to supply defects, and mechanical contrivances to aid physical inefficiencies—obliged by its construction to operate in the mass, or, if susceptible of a partial movement, only rendered so through the intervention of the most cumbrous and complicated machinery, constantly liable to disorder, and involving irremediable destruction to the whole from the slightest injury or interruption to the action of any one of its parts.

It is not our intention here to enter into an anatomical review of the conformation of the animal, or minutely to investigate the circumstances to which it is indebted for the peculiar aptitude we have stated it to possess for the element in which nature has destined it to move. The particular characteristics of its physical organization, the cavernous construction of its bones, by means of which the greatest strength is obtained with the least amount of weight, the immense muscular powers and mechanical advantages by and under which the wings are made to operate, the number and offices of the different nerves by which the influence of the will is conveyed to the minutest and most distant parts of the system, enabling it to direct its movements with equal efficacy and despatch, the peculiar perfection of the materials of which its appropriate organs are composed, uniting the strength and tenacity of the metals with the lightness and elasticity belonging to themselves alone ; and, lastly, the propriety of its own form, and that of all its parts, calculated to afford the least opposition where least is required, (as the head and chest in cleaving his way through the air, and the back of the wing in the act of recovering itself from the effects of its impact), and, on the contrary, displaying an active surface of resistance of the most efficient description where such a result is required of it; all these are peculiarities by which the bird is essentially enabled to secure the fulfilment of its purposes, but which would far exceed our present limits to discuss with any thing like the consideration they deserve.

From the difficulty that has been ever found to attend the first efforts of men to quit the ground by the aid of wings, and which seems to be experienced in some degree by all the larger specimens of the feathered tribe, a general opinion prevails that all birds in the commencement of their flight are obliged to resort to some adventitious expedient, as, for instance, springing from the ground by means of their legs, in order to attain a sufficient elevation preparatory to the exercise of their wings. As a general rule, inferring the insufficiency of the appropriate organs to the execution of their task, this supposition I take to be erroneous; first, because the smaller birds, who are chiefly presumed to practise it, have little or no need of the expedient, while the larger birds, to whom it might be thought more available, do not resort to it; and, secondly, because, if such an effort be necessary, aquatic birds, or those which occasionally frequent the water, would never be able to rise at all; the medium affording no sufficient opposition to enable them to give effect to the presumed extension of the legs. The difficulty in question is evidently one arising from the length of the wings, whereby their motion is obstructed by collision with the earth before sufficient resistance has been generated by their impact to effectuate the elevation of the body in the air, and is, consequently, liable to be experienced only by birds whose organs exceed the ordinary dimensions of their kind. In such cases the bird endeavours to obtain by the inclination of his ascent, the scope which is denied to his vertical exertions, and hence that running along the ground previous to the ascent which has been generally construed into a temporary inability to rise.

Having succeeded in getting clear of the ground, the next part of the operation requiring notice is the direction of the flight, which is effected by means of the same members in conjunction with another, the tail, the operation of which has been the subject of much speculation among natural historians, and even yet seems to be in general but very imperfectly understood. According to Ray, Willoughby, and many others, this organ is supposed to have the same effect upon the course of a bird, as a rudder has upon that of a ship; a very inconsiderate notion, for three reasons—first, because, by its horizontal position, it is especially incapacitated from fulfilling such an office; secondly, because it is much too small by its mere passive resistance to accomplish those rapid evolutions to which birds are particularly prone; and, thirdly, because such an appendage for such a purpose would be superfluous to the direction of a body, moving in a single medium, and impelled by equal forces on either side, or which can be rendered unequal by the volition of the animal itself. With equal inconsistency and incorrectness have others assigned to the tail the office of a counterpoise to enable the bird to maintain its equilibrium in the different dispositions of its centre of gravity when resting upon its wings and legs; an office for which it is as equally unfitted as the former; first, by reason of its weight, which is far too little to afford any compensation to the effects of so great a change in the position of the axis of equilibrium as must occur when the points of suspension are transferred from the one set of members to the other; and, secondly, because of its fixed position by means of which it is impossible that it could ever serve to maintain the equilibrium of a body, destined to change its axis of suspension, in both of its positions; for, if right in one case, it must evidently be wrong in the other. Add to this the fact that such a compensation as is here supposed is totally unnecessary; when standing, the bird brings its centre of gravity within its base, by the upright position of its body; and, when flying, it is no matter how its centre of gravity be disposed, as he is suspended from above. Lastly, it is supposed by some that the tail of a bird is designed to retard its descent after the manner of a parachute, in conjunction with the wings in a state of expansion; an absurd and unnecessary duty; for surely the power which is sufficient to enable the bird to rise against the attraction of gravitation, is sufficient to enable it to come down again with, at any rate, equal facility. The fact is, the operation of the tail of the bird is, in one sense of the word, clearly that of a *rudder;* not by occasioning the *lateral* inflexions of its course, but by contributing to its *vertical* motions; a modification which such (or some similar) agent is absolutely necessary to enable it to accomplish. This it does by the inclination it makes with the plane of the motion of the wings and the resistance it affords to the progress of the body in that direction to which these forces would of themselves impel it; whereby the fore-part of the body of the bird becomes more or less directed upwards, and the whole induced to assume a course generated by the resolution of the two. Without his tail the bird would have always to rise and fall perpendicularly (as concerns the posture of his body) and, except in a horizontal flight, would never be able to enjoy the advantage of opposing his natural *prow* to the resistance of the air. With regard to the horizontal divergence of its course, this is entirely effected by the action of the wing on the side opposed to that to which it designs to proceed; neither is any other required to second its endeavours.

It has been sought by most ornithologists to identify the course pursued by a bird in ascending or descending with some particular line, and the curve of a parabola has been fixed upon as the most convenient type for the purpose. Independently of the absurdity of assigning any mathematical figure to the determination of voluntary motion, and binding the free agents of God's creation to forms dictated by the laws of inanimate and insentient matter, the presumption is altogether inconsistent with reason or experience. The line which all birds would follow in proceeding directly from one object to another is a right line, were there no reasons for desiring an occasional deflexion. When a bird, therefore, starts from the top of a house or tree, with a view to reach any particular spot on the ground, it proceeds as directly as its natural mechanism will allow, until arriving near the ground, and, desirous to avoid concussion, it inclines its course a little towards the horizontal plane, and skims along till it reaches the object of its aim. The course it pursues in all these cases will certainly depend upon the relation between the point it starts from and that to which it tends, the rate it inclines to move at, and the particular structure of its motive organs. Where no impediment interposes to incline it to act otherwise, it will take a course as short as its structure will allow; as may be observed in watching the ascending motions of birds, when no solid object appears in the way, as in birds that soar; or, where there are such, when their position does not oblige them to fear a concussion in alighting; as in birds mounting to perch, or otherwise springing to attain a more elevated situation. Who can venture to assert that the lark in its early efforts to greet the dawn of day, the eagle in ascending to survey the subjacent plains in search of food, the partridge in its endeavours to escape the shot of the sportsman, or the falcon in making its deadly stoop upon the flock beneath it, affect in their course the curve of a parabola, or obey, indeed, any but that which is dictated by the spur of the moment, or the motions of the object of which it is in pursuit?

With regard to the portion of the sky, which may be said to be the proper sphere of the feathered tribe, the flight of birds, however exalted it may appear to the denizens of an humbler plane, is never really conducted at any considerable elevation above the surface of the earth. The utmost aspirations of the lark, the falcon, or the eagle, when soaring in mid heaven, as we are wont poetically to describe, never, I am inclined to believe, lead them to exceed an altitude of a few hundred feet above the level of the soil immediately beneath them. The ordinary flight of birds, however, is much lower, and, in fact, when viewed from the car of a balloon, even at a very moderate elevation, scarcely appears to merit the name of flight; so little does it seem to be removed from the contiguity of the neighbouring plains. In this respect, however, they but obey the course to which nature has particularly adapted them. It is not in the higher regions of the atmosphere that either their natural pleasures or natural wants can be most readily satisfied; while their removal from the means of shelter would render them an easier prey to those of their own kind that profit by their destruction, and at the same time expose them to the influence of atmospheric currents, with which they might not perhaps be able without difficulty to compete.

That this is the real cause of their confinement to an humble level, and not any incapacity on the part of the animal to range at higher levels, appears from the conduct of birds when occasionally transported to excessive elevations in balloons for experimental purposes, though some indeed would fain persuade us to the contrary. Mr. Green in the course of his practice has very frequently taken up with him pigeons, and other birds of various kinds, for the purpose of liberating them at a great altitude, and has never found any reason to suspect that they suffered in the least as to the motive energies from the reduction in the density of the medium. On one occasion he let loose a pigeon at an elevation of about 6000 feet; the earth being then totally excluded from view by a dense impenetrable layer of intervening clouds: the bird after quitting the car, ignorant where to proceed, (never perhaps having been placed in a similar situation) continued to accompany the balloon for several hours, maintaining the same distance, and traversing the empty hemisphere with idle circles, until the shades of evening and his own descent finally excluded it from his view. Upon another occasion Mr. Green took up with him a tame goose, (a bird not generally versed in the art of flying) and at a considerable elevation committed him to the exercise of his own resources. At first the bird, expressing a little proper astonishment at his novel situation, seemed dis. inclined to part company; being however at last persuaded to shift for himself, he commenced flying downwards, and so continued for about half a minute, till at length perceiving the inutility of labouring to do that which could be effected without it, he desisted from his exertions, and expanding his wings, allowed himself to descend in gradually increasing circles to the ground. M. Gay-Lussac, however, in his first voyage, in company with M. Biot, exhibited the fact in a still stronger light, by dismissing a linnet from the car at a height of 11,000 feet, which, except in the circumstance of having directed its flight towards the earth, discovered no unusual symptoms of uneasiness or incapacity either in its conduct or appearance.

But what need we the aid of aerostation to determine a fact of which we have the most conclusive testimony in the natural flight of the condor, one of the greatest and heaviest of the feathered tribe, above the lofty peaks of the Andes, many thousand feet beyond the highest point to which any, for the purposes of experiment, have ever been artificially transported ? To fear, therefore, and confusion at the novelty of the situation, together with the cramped and awkward position in which they have been previously confined, are no doubt to be attributed those symptoms of weakness and hesitation which some aeronauts have observed in the repetition of these experiments, and have inconsiderately interpreted into an inability to sustain themselves, and a fear to trust to their own exertions in a medium which the excessive elevation had rendered unsuited to their support. What might be their condition at still higher elevations remains to be ascertained.

APPENDIX G.

THE following stanzas have been selected out of a
large number on the same subject, which were presented
to us or circulated during our stay at Weilburg, and will
be found to form a very creditable commentary upon the
state of the arts and literature in that comparatively
secluded part of the continent. I have been chiefly
induced to repeat them here as being calculated to
exemplify in terms of their own dictation the very kindly
feelings that universally prevailed in regard of our unexpected visit, and the very gratifying intercourse to
which it gave rise. As effusions in the dead languages
may not be quite as intelligible to ordinary readers, as
in the living, we have subjoined to each as tolerable a
translation as our poor abilities would enable us to
indite.

No. I. is from the pen of FRIEDERICH TRAUGOTT
FRIEDEMANN, Doctor of Theology and Philosophy in
the College (termed the Gymnasium) at Weilburg, and

was addressed to Mr. Green, on the occasion of a fête given to us by the principal inhabitants of the town, the evening before our departure.

CAROLO GREENIO,

Britanno, Artifici peritissimo et clarissimo.

> " Expertus vacuum Dædalus aëra
>   Pennis non homini datis.
>   Nil mortalibus arduum est.
>   Cælum ipsum petimus."
>
>   HORAT.

OLIM *Blanchardus,* nostras delapsus ad oras,
  Insolitâ clarum nomen ab arte tulit.
Sed breve per spatium, Mæno Taunoque relicto,
  Finis erat celeris Vilinaburga viæ.
Nunc, post lustra decem, majus quid Greenius ausus,
  A Tamesi ad Lanam deproperavit iter.
Per mare, per fluvios, montes transgressus et urbes,
  Nec noctis tenebras horruit impavidus.
Cedite, Romani, Graii quoque, cedite Galli;
  Dædaleam laurum Greenius uuus habet.

TO CHARLES GREEN, &c.

First Dædalus the empyrean tried,
  With wings which ne'er to man were given.
Nought is too great for mortal stride.
  We seek the very gates of Heaven.

In former days the Gallic Blanchard came
To Weilburg's humble shores, in search of fame.
But short his flight, and brief his bold career,
From Francfort and the Maine transported here.

But who is he that now, with bolder span,
Flies from the Thames to settle on the Lahn,*
O'erleaping empires in his daring flight,
Regardless of the sable frowns of night?
Yield, Romans; yield, Greeks; Frenchmen quit the throne;
The crown of Dædalus is Green's alone.

No. II.   The following valedictory address is by
HERR APPAL, and was likewise presented to us on the
before-mentioned occasion.

Heu! heu! pennatis zephyris quid fertur in alto
Æthere, non visum prodigium pavidis?
Num superi cœlo descendunt, sedibus altis,
Ut quos in terris destituere petant?
Num facies diræ, vastum trans æthera missæ,
Delicta ulturæ, quæ dedit impia mens?
Tres sunt audaces magno molimine nisi,
Regna Jovis magni quærere fulminea.
Anglia, clara opibus, tulit atque Hibernia tellus,
Terram Nassoicam qui petiere Noto.
Mane erat.   In nimbis clarus conspicitur orbis
Grandis, subque pila parva subnexa ratis.
Ruricolæ fugiunt, horrendum dæmona cœlo
Delapsum esse rati, Tartareamque luem.
Anchora de navi jacitur; stat machina corupis;
Terram exoptatam tangere perque juvat.
Fama volat cursu celeri, qui nuntiet urbi,
Devenisse viros, quique quibusque locis.
Apparent illi, magnâ comitante catervâ,
Et sibi cunctorum concilient animos.

* The river on which is seated the town of Weilburg.

Jam vale tu Greeni, vale Masoni, vale tu Hollond;
 Et faveant vestris cœlicolæ meritis :
Atque hinc regressi in patriam, vestrosque penates,
Nomen Weilburgi pectore sit memori.

Look ! look ! what fearful prodigy is there,
Riding upon the bosom of the air ?
Can they be gods descended from above,
In hopes the Heaven they've left, on earth to prove ?
Or are they demons from a deadlier clime,
Sent to exact the penalty of crime ?
Three dauntless men they are, who borne on high,
Invade the vast dominion of the sky.
From Britain's fertile plains—their native shore—
They come the land of Nassau to explore.
'Twas morn ; when lo ! a glitt'ring orb appears ;
High o'er the clouds its head majestic rears.
The affrighted peasants fly, deeming at hand
Some mighty portent sent to scourge the land.
The anchor's cast ; they rest upon the plain ;
And joyful tread the wished-for earth again.
Fame now with rapid speech the event proclaims,
And tells to all their nation and their names.
At last, the centre of a crowd, they come,
And courteous earn in every heart a home.

Now farewell Hollond, Mason, farewell Green !
May Heaven propitious to your worth be seen !
And when you to your native homes depart,
Be Weilburg's name engraved upon each heart.

No. III. The name of the author of the following having escaped my memory, all I can do is to record his sentiments.

Den HERREN KARL GREEN, ROBERT HOLLOND, UND THOMAS MONCK MASON,

AUS ENGLAND,

Bei ihrer abreise von Weilburg, Am 19 November, 1836.

———

LONDON.

Der Geist, des Zugs der Himmelskraft bewusst,
Folgt seinem Ruf zu unermess'nen Fernen,
Sucht Wohnung sich in glanzdurchwobenen Sternen,
Zu stillen seines Sehnens tiefe Lust.

Er bricht die Fesseln seiner irdischen Kraft,
Erweitert freudig kühn des Wissens Räume—
Weit unter sich die goldigen Wolkensäume—
Ein Dädalus, der selbst sich Flügel schafft.

So glühet Ihr, ein würdiges Ziel erstrebend,
Das hehre Eigenthum der Sterblichen zu weiten;
Der Kunst gilt Euer muthiger Entschluss!

Die Brust zu edlen Hoffnungen erhebend,
Entschwebt Ihr zu des Aethers reinern Freuden
Und jubelnd schallt Euch nach ner Freunde Gruss!

DIE LUFT.

Glück auf! Entrückt dem niederen gewühl,
Ihr schiffet in des blauen Himmels Wellen,
Wo Hochsinn, Freudigkeit die Herzen schwellen,
Und schwelget reich in seligem gefühl.

Fern von des Eigennutzes argem Streit
Kennt Ihr die Liebe nur, die Euch verbündet,
Und die in grüssen Ihr nach unten kündet,
In werthen Zeichen Eurer Freundlichkeit.

Und ob in rein're Sphären auch gehoben,
Wahrt Ihr doch treuen Sinn und blickt von oben
Mit Sehnsucht nach dem Mutterboden hin.

So zieht's den weisen, dessen edles Leben
Sich hoher Wissenschaft und kunst ergeben,
Doch immer wieder zu den Brüdern hin !

WEILBURG.

Willkommen denn ! auf Nassau's deutschem Grund
Willkommen, freundlichste der Gäste,
Wir bieten von dem Unsern Euch das Beste;
So spricht das Herz, nicht nur der Mund.

Dank Eurer Güte, die durch That und Wort
Sich hat ein dauernd Monument begründet ;
Den Kranz der Dankbarkeit Euch nicht entwindet
Die ferne Zeit ; er grünet Euch hier fort !

Schon wollt' Ihr wieder lassen unsre Mauern,
O glaubt es, alle Bürger Weilburg's trauern,
Ihr Lebewohl ist herzlich wahr und rein.

Denkt unser auch im fernen Vaterlande,
Nicht lös die schönen Freundschaftsbande,
Hier sollet Ihr stets unvergessen sein.

THE END.